# The Environmental Debate

# THE
# ENVIRONMENTAL
# DEBATE

*A Documentary History*

Edited by PENINAH NEIMARK
and PETER RHOADES MOTT

Primary Documents in American History and Contemporary Issues

GREENWOOD PRESS
Westport, Connecticut • London

**Library of Congress Cataloging-in-Publication Data**

The environmental debate : a documentary history / edited by Peninah
Neimark and Peter Rhoades Mott.
    p.  cm.—(Primary documents in American history and
contemporary issues, ISSN 1069–5605)
  Includes bibliographical references and index.
  ISBN 0–313–30020–8 (alk. paper)
    1. Environmentalism—United States—History.  I. Neimark,
Peninah, 1938–  .  II. Mott, Peter Rhoades.  III. Series.
GE197.E7    1999
363.7'05'0973—dc21      99–17844

British Library Cataloguing in Publication Data is available.

Library of Congress Catalog Card Number: 99–17844
ISBN: 0–313–30020–8
ISSN: 1069–5605

First published in 1999

Greenwood Press, 88 Post Road West, Westport, CT 06881
An imprint of Greenwood Publishing Group, Inc.
www.greenwood.com

Printed in the United States of America

The paper used in this book complies with the
Permanent Paper Standard issued by the National
Information Standards Organization (Z39.48–1984).

10 9 8 7 6 5 4 3 2 1

## Copyright Acknowledgments

Every reasonable effort has been made to trace the owners of copyright materials in this book, but in some instances this has proven impossible. The editors and publisher will be glad to receive information leading to more complete acknowledgments in subsequent printings of the book and in the meantime extend their apologies for any omissions.

The editors and publisher are grateful to the following for granting permission to reprint excerpts from their materials:

Document 79: 100,000,000 GUINEA PIGS by Arthur Kallet and F. J. Schlink. Copyright© 1933, 1960 by Arthur Kallet and F. J. Schlink. Reprinted by permission of Vanguard Press, a division of Random House, Inc.

Document 81: Arthur Tansley, "The Use and Abuse of Vegetational Concepts and Terms," *Ecology* 16 (1935), courtesy of the Ecological Society of America.

Document 83: THE CULTURE OF CITIES by Lewis Mumford, copyright 1938 by Harcourt Brace & Company and renewed 1966 by Lewis Mumford, reprinted by permission of the publisher and Martin Secker & Warburg, Ltd.

Document 84: THE GRAPES OF WRATH by John Steinbeck. Copyright 1939, renewed© 1967 by John Steinbeck. Used by permission of Viking Penguin, a division of Penguin Books USA Inc. Copyright© 1939 by John Steinbeck, reprinted by permission of McIntosh and Otis, Inc., and William Heinemann, publisher.

Document 85: Marjory Stoneman Douglas, *The Everglades: River of Grass* (1947), courtesy of University of Miami, Archives and Special Collection Division, Otto G. Richter Library, Coral Gables, Florida.

Document 86: Roger Tony Peterson, *Birds over America* (New York: Dodd, Mead, 1948), pp. 71–73, by permission of Mrs. Virginia Peterson.

Document 87: OUR PLUNDERED PLANET by Fairfield Osborn. Copyright 1948 by Fairfield Osborn. By permission of Little, Brown and Company.

Document 88: A SAND COUNTY ALMANAC: AND SKETCHES HERE AND THERE, SPECIAL COMMEMORATIVE EDITION by Aldo Leopold. Copyright 1949, 1977 by Oxford University Press, Inc. Used by permission of Oxford University Press, Inc.

Document 94: John Kenneth Galbraith, "How Much Should a Country Consume?" in Henry Jarrett, ed., *Perspectives on Conservation: Essays on America's Natural Resources*, pp. 98–99, © 1958. The Johns Hopkins University Press.

Document 95: David Brower, *For Earth's Sake* (Salt Lake City: Peregrine Smith, 1990), courtesy of David Brower.

Document 97: Lorus J. Milne, Ph.D., and Margery Milne, Ph.D., *The Balance of Nature* (New York: Knopf, 1961), courtesy of Margery Milne, Ph.D.

Document 98: Lewis Herber (Murray Bookchin), *Our Synthetic Environment* (New York: Knopf, 1962), courtesy of Murray Bookchin.

Document 99: "A Fable for Tomorrow," SILENT SPRING by Rachel Carson. Copyright© 1962 by Rachel L. Carson. Copyright© renewed 1990 by Roger Christie. Reprinted by permission of Houghton Mifflin Company and Frances Collin, Trustee. All rights reserved.

Document 100: Stewart L. Udall, *The Quiet Crisis* (New York: Holt, Rinehart and Winston, 1963), courtesy of Stewart L. Udall.

# Contents

# Series Foreword

This series is designed to meet the research needs of high school and college students by making available in one volume the key primary documents on a given historical event or contemporary issue. Documents include speeches and letters, congressional testimony, Supreme Court and lower court decisions, government reports, biographical accounts, position papers, statutes, and news stories.

The purpose of the series is twofold: (1) to provide substantive and background material on an event or issue through the texts of pivotal primary documents that shaped policy or law, raised controversy, or influenced the course of events, and (2) to trace the controversial aspects of the event or issue through documents that represent a variety of viewpoints. Documents for each volume have been selected by a recognized specialist in that subject with the advice of a board of other subject specialists, school librarians, and teachers.

To place the subject in historical perspective, the volume editor has prepared an introductory overview and a chronology of events. Documents are organized either chronologically or topically. The documents are full text or, if unusually long, have been excerpted by the volume editor. To facilitate understanding, each document is accompanied by an explanatory introduction.

It is the hope of Greenwood Press that this series will enable students and other readers to use primary documents more easily in their research, to exercise critical thinking skills by examining the key documents in American history and public policy, and to critique the variety of viewpoints represented by this selection of documents.

# Preface

The technological innovations that produced, first, the Agricultural Revolution, then, the Industrial Revolution, and, in the latter part of the twentieth century, the Information Revolution have enabled people not only to manipulate the natural environment, but to do so on an ever grander scale and at an ever accelerating pace. The use of biotechnology and other new technologies undoubtedly will enable people to make even more extensive and more rapid changes in their environment in the future.

As long as there existed on earth uninhabited or sparsely inhabited lands and as long as the impact of human technological innovation on the environment was confined to small segments of the planet, the environmental consequences of the human use of technology and human population growth could be largely ignored. Today, though, there are few frontiers left on earth, and many human activities have global impacts. Chemicals released into the atmosphere cause acid rain that destroys forests hundreds of miles away from the point of source and may even affect the climate of the whole earth; the fishing and ocean dumping practices of one nation impinge on the availability of seafood to nations around the globe; microbes on food produced in one country result in outbreaks of illness thousands of miles away; people caught in the crossfire of war or facing environmental degradation in their homelands may emigrate extremely rapidly en masse to very distant parts of the world. Hence, Americans, as well as other people, are being forced to confront the limitations of their environment—the earth.

During the coming decades, people throughout the world will have to make very difficult economic and social choices as they attempt to balance human needs with environmental limitations. They will have to decide in what kinds of technologies their nation should invest (e.g.,

alternative energy sources, desalinization), what kinds of limits on individual rights they are willing to accept (e.g., limits on private property
rights, limits on child-bearing rights), and on what basis they want to
make these choices (e.g., economic impact, ethics).

Recognition that these choices will affect how we and future generations of Americans will live has prompted vociferous arguments about
where our nation's priorities should lie. As we gathered the documents
for this book, we became increasingly aware that although there has
always been a broad spectrum of opinions about how to deal with resource and environmental issues, the most strident voices at the two
extremes of the environmental debate often have been the most effective
in raising people's consciousness about these issues. At one extreme are
those who argue that we must immediately begin diminishing our impact on the environment and give greater priority to the rights of the
natural world; at the other are those who argue that we should use all
available resources to further human well-being and that human inge-
nuity will allow us to continue to do so. The debate between these two
sides has prodded people in the United States to recognize not only that
economic well-being is dependent on environmental well-being but also
that pristine wilderness areas and unspoiled rivers will not continue to
exist if the people who live near them and depend on them for their
livelihoods are in economic distress. Only by listening to all the voices
in the environmental debate can we begin to comprehend the long-term
social, economic, and political ramifications of alternative solutions to
complex environmental problems.

The documents in this book chronicle the evolution of environmental
movements in the United States and show how environmental interests
and organizations have evolved alongside changing social, economic,
and political conditions and changing American views about the environment. They reveal patterns of changing attitudes about three
fundamental environmental issues: conservation, pollution, and the
human-nature relationship.

Selections concerned with conservation focus on land and resource use
and development, and record the change from a belief that America's
resources are virtually unlimited to an understanding that particular resources are finite and that how we use one resource can affect the availability of other resources. These documents offer an overview of how
Americans have thought about the natural resources of this country and
how they have dealt with conflicts stemming from the need to use these
resources and the desire to guarantee their availability for future generations. Selections relating to pollution focus on air and water quality,
as well as toxic chemicals and waste disposal, and they document an
evolution from minimal awareness of the effects of pollution to serious
concern and demands that people and the government take action. Se-

lections centered on the human-nature relationship deal with the issues of the preservation of open space and wilderness areas and the protection of species from extinction. Here we find a progression from an anthropocentric (human-centered) view of the earth to a more biocentric (all-living-organisms-centered) understanding. The writings depict efforts to turn people from aggressive predators into thoughtful custodians of an environment shared with other living things and underscore an increasing desire of humans to have closer contact with nature than ordinarily may be available for them.

The documents give evidence of the continuing interaction between environmental factors and the course of development in the United States. They detail the growth of concern about the connections among increasing numbers of people, expanding industrialization, technological change, urbanization, pollution, and resource depletion. Although few of the documents deal directly with technological development, several of the more recent documents underline the need to recognize limits to our ability to control nature.

In selecting material for inclusion, we have sought writings and official documents that indicated very early interest in particular environmental issues or showed remarkable foresight into problems that would later become serious environmental concerns, writings that had a major impact on the development of environmental concern and policy, and writings that captured the tenor of the thinking of a particular period. We have also tried to show the evolution of both popular attitudes and government policy concerning the environment and the use of natural resources. While attempting to cover a broad range of environmental concerns and movements, we have also focused on a few areas of environmental concern—including land development, use of resources, waste disposal, wetlands, and wildlife—and closely followed how the evolution of public attitudes has affected government policy in these areas.

We have included selections from public documents, letters, speeches, and essays, as well as poetry and fiction that offer a wide array of attitudes about environmental issues. This book presents the views of those who have recognized that natural resources are finite and yet believe that it is possible to conserve them or limit their use; of those who have thought that people should take as much from the earth as they wish, believing that technological know-how will always enable humanity to overcome resource limitations; and of those who have been convinced that human activity must take into account the rights of other living things. The writers include naturalists, among them botanists and ornithologists; conservationists, ranging from forest managers to game hunters to grass-roots activists; scientists, philosophers, and theologians; lawyers and judges; politicians and industrialists; historians, sociologists, and economists; artists, architects, and designers; and poets and novel-ists. As the selections indicate, while environmental action generally stems from concern about a particular local issue—pollution in Plymouth Colony harbor, smog in Los Angeles, the decline in the osprey popula-tion on Long Island, New York—since the nineteenth century, activists have recognized that local remedies frequently are inadequate, and that they must effect state, federal, or even international change if they want to obtain satisfactory solutions.

# Acknowledgments

We thank the many people who have assisted us, directly or indirectly, in the process of preparing this book, providing us with suggestions for selections and helping us to locate hard-to-find documents. First, we acknowledge the guidance offered by the editors of three earlier documentary histories of environmentalism: Robert McHenry and Charles Van Doren (*A Documentary History of Conservation in America*), Roderick Nash (*Readings in the History of Conservation*), and Carolyn Merchant (*Major Problems in American Environmental History*). In their books we found not only ideas for documents to include but also evidence of the need to cover certain issues and present particular viewpoints.

We are grateful for the help we received from librarians and other staff members at various libraries, archives, and historical societies, including Liz Burns at the Brookhaven Public Library in Brookhaven, New York; Kenneth Carlson at the Rhode Island State Archives in Providence; Grace Korbutt at the Plymouth Public Library in Plymouth, Massachusetts; Cornelia Locher at the Tate Library, Fieldston School, Bronx, New York; and Daryl Morrison, head of the John Muir Collection at the University of the Pacific, Stockton, California.

Our advisers—Timothy Dantoin of Environmental Technology Review; Peter Lehner, an attorney for the Natural Resources Defense Council and the New York Attorney General's Office, Environmental Protection Bureau; and Susan Senecah of the College of Environmental Science and Forestry of the State University of New York at Syracuse—not only brought to our attention several of the documents that we included but also made numerous suggestions about topics that merited discussion in the text. We especially owe Susan thanks for her thoughtful critiques of parts of early versions of the manuscript.

We are indebted to Frances Beinecke and Mark Izeman of the Natural

Resources Defense Council and Albert Butzel for supplying us with copies of key U.S. legal documents, to Jim Sniffen of the United Nations Environmental Programme and Howard Schiffman for providing us with copies of international agreements relating to the environment and for assistance in preparing the list of major international environmental agreements, and to David Helvarg for directing us to useful source materials. To Leonore Hauck, who has served as a mentor to Nina in her career as an editor, and to Bob Montera, we owe thanks for their review of sections of the manuscript.

We also thank our spouses, Harold Neimark and Lenore Mott, and our children, Geoff Neimark and Peter Mott—Harold for bringing to our attention numerous references in the scientific literature and for being especially supportive at those times when the scope of the project and the mass of material confronting us seemed overwhelming; Lenore for being understanding about the weekends and evenings used to work on this book; Geoff for his assistance in locating documents and for his patient instructions on the use of new computer software; and Peter for being readily available for consultation about computer problems.

We are grateful to Beverly Miller for her careful copyediting of the text, and Betty Pessagno for overseeing the production of the book. Finally, we thank Emily Michie Birch, our editor at Greenwood, for her encouragement during the three years that it took us to complete this book, and Penny Sippel for suggesting that we undertake the project.

# Significant Dates in American Environmental History

| | |
|---|---|
| 1626 | Plymouth Colony passes ordinance regulating cutting and selling of timber. |
| 1637 | Plymouth Colony passes ordinance regulating herring run. |
| 1639 | Newport, Rhode Island, prohibits deer hunting for six months of the year. |
| 1681 | William Penn, proprietor of Pennsylvania, requires that one acre of land be left forested for every five acres of land cleared. |
| 1691 | Charter renewal for Massachusetts Bay establishes Broad Arrow Policy, setting aside large trees suitable for ship masts by marking them with a "broad arrow." |
| 1692 | American Philosophical Society formed to promote useful knowledge and encourage scientific agriculture. |
| 1804–1806 | Lewis and Clark expedition undertakes first federal survey of nation's resources. |
| 1828–1831 | First attempt at federal forest management undertaken on Santa Rosa Island, Florida. |
| 1832 | George Catlin proposes a national park. |
| 1844 | William Cullen Bryant proposes a great municipal park. |
| 1848 | Gold discovered at Sutter's Mill, California; marks the beginning of the California gold rush. |
| | American Association for the Advancement of Science founded. |
| 1849 | U.S. Department of the Interior established and given authority to administer public lands. |
| 1864 | Yosemite Valley, California, set aside as a state park. |
| | George Perkins Marsh publishes *Man and Nature*. |

| | |
|---|---|
| 1865 | Work begins on Central Park in New York City, the nation's first "rural" city park. |
| 1869 | John Wesley Powell descends the Colorado River and travels through the Grand Canyon. |
| 1871 | U.S. Fish Commission established. |
| 1872 | April 10 designated as Arbor Day as a result of a campaign spearheaded by J. Sterling Morton. |
| | Yellowstone National Park established as first national park in the world. |
| 1875 | American Forestry Association organized to promote forestry. |
| 1876 | Franklin B. Hough appointed special agent in Department of Agriculture to study forest conditions. |
| | Appalachian Mountain Club founded. |
| 1879 | U.S. Geological Survey established. |
| | National Board of Health formed. |
| 1881 | Division of Forestry created within the Department of Agriculture for the purpose of fact finding. |
| 1885 | Niagara Reservation created to protect Niagara Falls through the cooperative effort of New York State and the Canadian province of Ontario. |
| | Adirondack Forest Preserve (later, Adirondack State Park) set aside in New York State. |
| | The Division of Economic Ornithology and Mammology (predecessor of the U.S. Biological Survey) established within the Department of Agriculture. |
| 1886 | Audubon Society founded by George Bird Grinnell. |
| | Mission of the Division of Forestry is expanded and Bernhard E. Fernow is named division director. |
| 1887 | Boone and Crockett Club founded. |
| 1891 | President Benjamin Harrison sets aside 13 million forested acres as forest preserves. |
| | Yosemite National Park established. |
| 1892 | Sierra Club organized by John Muir. |
| 1895 | American Scenic and Historic Preservation Society founded. |
| 1896 | Hawaiian koa finch becomes extinct. |
| 1898 | Cornell University offers first college-level course in forestry. |
| | Gifford Pinchot becomes head of the Division of Forestry. |
| | Wildlife Conservation International formed. |
| 1902 | Bureau of Reclamation established within the Department of the Interior. |

| | |
|---|---|
| 1905 | National Audubon Society formed. |
| | Responsibility for forest reserves transferred from the Department of the Interior to the Department of Agriculture's new Forest Service. |
| | Pelican Island, Florida, set aside by President Theodore Roosevelt as first national wildlife refuge, with the goal of protecting brown pelican nesting sites. |
| 1907 | Inland Waterways Commission established. |
| 1908 | Grand Canyon named a national monument. |
| | Governors' conference on conservation convened by Theodore Roosevelt. |
| | National Conservation Commission appointed to inventory the nation's natural resources. |
| 1909 | North American Conservation Conference convened by Theodore Roosevelt. |
| 1909–1910 | Controversy between Richard Ballinger and Gifford Pinchot disrupts federal conservation activities. |
| 1910 | Bureau of Mines established within the Department of the Interior. |
| 1911 | American Game Protective and Propagation Association established. |
| 1913 | Hetch Hetchy Valley in Yosemite National Park given to San Francisco for building a reservoir, ending John Muir–led campaign to preserve the valley. |
| 1914 | Last passenger pigeon and Carolina parakeet die. |
| 1915 | Ecological Society of America founded. |
| 1916 | National Parks Service established. |
| 1919 | National Parks and Conservation Association established. |
| 1920 | Federal Power Commission established. |
| 1922 | Izaak Walton League of America founded. |
| | *Pennsylvania Coal Company v. Mahon et al.* raises issue of the right of government to restrict use of private land without compensation to owners. |
| 1924 | Teapot Dome scandal. |
| | Gila National Forest, New Mexico, designated as the first extensive wilderness area by the Forest Service. |
| 1926 | *Village of Euclid v. Ambler Realty Co.* establishes that local governments may enact zoning laws that limit the right to use private property. |

| | |
|---|---|
| 1928 | Boulder Canyon Project (Hoover Dam) authorized, with the goal of expanding flood control, irrigation, and electrification in western states. |
| 1933 | Civilian Conservation Corps established. |
| | Tennessee Valley Authority created. |
| 1935 | Wilderness Society organized. |
| | Soil Conservation Service created within the Department of Agriculture. |
| 1936 | National Wildlife Federation founded. |
| 1937 | Ducks Unlimited founded. |
| 1940 | U.S. Fish and Wildlife Service created. |
| 1945 | United States drops atomic bomb on Hiroshima, Japan. |
| 1946 | Atomic Energy Commission established. |
| | U.S. Bureau of Land Management created to consolidate management of lands in the public domain. |
| 1947 | Conservation Foundation organized. |
| | Defenders of Wildlife organized. |
| 1948 | Heavy smog in Donora, Pennsylvania, causes several deaths. |
| 1949 | First Sierra Club Biennial Wilderness Conference held. |
| 1950 | New Mexican grizzly bear becomes extinct. |
| 1951 | Nature Conservancy established. |
| 1952 | President's Materials Policy Commission issues report. |
| 1955 | Resources for the Future formed to study U.S. resource use. |
| 1956 | Proposed Echo Park Dam, planned for construction in Dinosaur National Monument, deleted from Upper Colorado River Storage Project; marks a major victory for wilderness preservation. |
| 1958 | Outdoor Recreation Resources Review Commission appointed by Congress. |
| 1959 | Trout Unlimited formed. |
| 1960 | Congress allocates funds for environmental health activities in the budget of the Department of Health, Education, and Welfare. |
| 1961 | World Wildlife Fund–U.S. established. |
| 1962 | Rachel Carson publishes *Silent Spring*. |
| 1963 | Dispute between Arizona and California over allocation of Colorado River water. |
| 1965 | *Scenic Hudson Preservation Conference v. Federal Power Commission* establishes that environmental factors as well as economic factors must be given consideration in the planning of federal construction projects. |

1967        Environmental Defense Fund established.

1968        California Air Resources Board created.

1969        Oil spill near Santa Barbara, California.

            Union of Concerned Scientists founded.

            Council on Environmental Quality created.

1970        National Oceanic and Atmospheric Administration established.

            Environmental Protection Agency established.

            Natural Resources Defense Council founded.

            First Earth Day celebrated on April 22.

1971        Greenpeace founded.

1972        *Sierra Club v. Morton* raises the issue of the rights of nature in a court of law.

            Oregon passes first state bottle recycling law.

            Environmental Protection Agency bans interstate sales of DDT.

            United Nations Conference on the Human Environment held in Stockholm; marks the beginning of the United Nations Environmental Programme (UNEP) and of a global approach to environmental problems.

1973        Oil embargo by the Organization of Petroleum Exporting Countries.

1974        World Watch Institute founded.

1977        Department of Energy formed.

1978        Chlorofluorocarbons banned from use in spray cans.

            Community of Love Canal, near Niagara, New York, evacuated because of presence of toxic wastes.

1979        Toxic chemical dumping practices of the Hooker-Occidental Chemicals Plant in Lathrop, California, become a public issue.

            Partial meltdown of nuclear reactor at Three Mile Island, near Harrisburg, Pennsylvania.

1980        National Acid Precipitation Program begun.

            Agency for Toxic Substances and Disease Registry established within the Public Health Service.

1982        World Resources Institute established, with headquarters in Washington, D.C.

1984        Toxic gas from a plant owned by Union Carbide, an American company, kills more than 2,000 people in Bhopal, India.

1985        Federal Conservation Reserve Program established to remove environmentally sensitive farmland from agricultural production.

            United Nations Conference on Ozone Depletion held in Vienna.

| 1989 | *Exxon Valdez* oil spill leaks 35,000 tons of oil into Prince William Sound in Alaska. |
| 1992 | United Nations Conference on Environment and Development held in Rio de Janeiro. |
| 1997 | United Nations Conference on Global Change held in Kyoto. |
| 1999 | Invasive Species Council established, bringing together the departments of Interior, Transportation, Agriculture, Commerce, and Defense and the Environmental Protection Agency to prevent and control invasive species. |

Seven species of Pacific salmon added to the Fish and Wildlife Service's list of endangered species, the largest regional group of species ever added simultaneously to the list.

Appeals court rules that standards for permissible levels of ozone and fine soot set by the Environmental Protection Agency in 1997 are invalid because in setting those standards the EPA exceeded the powers granted to it under the Clean Air Act.

# Introduction

Many of the environmental problems we face as we enter the twenty-first century are not new. Although the tremendous increases in population and the technological developments of the past fifty to seventy-five years have created some previously unknown problems, many current environmental problems are merely exacerbations of the kinds of problems that have arisen whenever significant numbers of humans have settled in a region for long periods.

Anthropologists think that environmental problems, stemming from too little water or land to support dense population centers, may have contributed to the collapse of various ancient civilizations, including the Anaszi and the Maya. In colonial America, interference with the spawning of fish caused by the damming of waterways to supply energy for mills and harbor pollution were causes of friction and concern. In the early nineteenth century, the issue of sufficient land to support massive population growth was already being discussed in the United States as well as in Europe. By the middle of the century, air pollution, water supply, and urban sanitation were well-recognized problems. During the same period, voices were being raised in opposition to the wanton destruction of wildlife and calling for the preservation of forests and scenic land. However, it was not until the 1890s, when the U.S. western frontier had disappeared and there was no longer open land for Americans to expand into, that conservation became a major issue. And it was only after the dropping of the atomic bomb in 1945 and the realization that people had the ability to make the earth uninhabitable that the concept of environmentalism gained massive popular support.

Although problems of resource depletion, pollution, environmental degradation, and population growth have long been recognized, finding acceptable solutions to these problems has rarely been simple because people expect the solutions to satisfy three fundamental human needs and desires: (1) material and economic well-being, including adequate food, clothing, and shelter; (2) physical well-being, including health and safety; and (3) access to open space and other aspects of nature that enhance spiritual well-being. Attempts to find solutions invariably arouse competing economic, social, and political interests and ethical and religious values.

## POPULATION

Environmental problems form a web, with the issue of population growth at its center. The larger the number of people, the more land that will be occupied, the more energy that will be used, the more resources that will be consumed, and the more waste that will be produced. How many people there are in a finite area, how they live, and how they use resources not only affect what happens to all living things in that particular area, but may also impinge on living things, including people, elsewhere on the planet.

The rate of human population growth often surprises those unfamiliar with population dynamics. It took until 1850 for the world population to reach 1 billion people and only another 80 years for it to reach 2 billion. Just 30 years later the earth contained 3 billion people. Between 1960 and 1975 the world's population grew by another billion, in 1987 it reached 5 billion,[1] and before the end of 1999 it reached 6 billion. While the rate of population growth in the United States and in most of the rest of the world has slowed in recent years, the actual numerical increase in population continues to be staggering. Each year about 86 million people are added to the world population, 98 percent of them in developing countries. In a few countries, including Japan and 30 European nations, population seems to have stabilized or is even declining, but in most countries, including the United States, it continues to increase. (In the United States the annual rate of population increase is 0.8 percent, the highest in the industrial world.)

Food supply, habitat suitability, disease, and predation limit the number of individual members of a particular species that can live in a given ecosystem. Humans, however, have developed technologies to manipulate and control these factors, and as their technological abilities have advanced, their population has been able to increase. Although there still are places where people die of starvation, at present there actually is sufficient food in the world to feed everyone; today, people starve only because political, social, or economic factors prevent them from obtaining food. Even though some places are unsuitable for human habitation, humans can now live almost any place on earth if they employ sufficient, and expensive, technology. As we produce food more efficiently, reduce death from disease and starvation, and improve nutrition and health care, we make it possible for more babies to be born, for more children to reach adulthood, and for the elderly to live longer. Nevertheless, it is predicted that sometime within the next few centuries, economic, political, and cultural factors will lead to a sufficient decline in the birth rate to bring about a leveling off of the world's population.

## LAND USE AND PROPERTY RIGHTS

Housing, agriculture, grazing, forests, mining, industry, and transportation all take up land, and the larger the population, the more land each of these purposes generally requires. However, since land that is used for one purpose usually cannot be used simultaneously for another purpose, population growth in a region or community tends to spur conflicts between individuals or groups who want to use the available land for different purposes.

The past few decades have witnessed increasing conflict between those who want to develop the land (including those who want to "reclaim" wet or arid land) and those who want to limit development, control urban sprawl, preserve open space and wilderness, and prevent the removal of land from agricultural use. Conflicts have also arisen when runoff from agricultural land has despoiled rivers and lakes and when the reclaiming of wetlands for farming or home sites has interfered with the land's normal ability to absorb waste and water.

As humans use land, less of it is available to the organisms that traditionally inhabited it. In some cases, land can be shared, but only rarely is human intrusion not followed by major disruption, displacement, or elimination of the area's native organisms. In addition, human use usually brings with it, wittingly or unwittingly, a host of exotic species that prosper in their new habitat and frequently drive out native species.

Because so many social, economic, and environmental issues are intertwined with land-use issues, efforts to find creative solutions to major land-use conflicts have stimulated people to reexamine some of their fundamental ideas about both property rights and the relationship between humans and the natural world.

## WATER AVAILABILITY AND WATER QUALITY

People must have an adequate supply of water, and it must be clean enough to use. However, since water resources—like land resources—are finite, expanding populations place ever greater stress on the available water both because of the greater number of people who need water and the increased human activity that results in the diversion and contamination of that water. In addition, the expanding use of toilets, bathtubs, showers, washing machines, dishwashers, waste disposals, and other appliances that employ water has led to a continually increasing per capita use of water.

Conflict over water use has existed for a long time in this country, and it will undoubtedly increase as cities and towns continue to expand. Furthermore, water for irrigation is the lifeblood of the economy in some agricultural areas, and the expansion of agriculture, especially in arid regions, increases the demand for water. As the demand for water increases, there may not be sufficient supplies of water in years of drought or in arid regions of the country to meet the demand. In those parts of the country with little annual rainfall, intense conflicts have already arisen over who should have access to the limited water supply: Should the bulk of it go to individuals, to agriculture, or to industry? In the Southwest, California and Arizona have been battling for many decades over allocation of water from the Colorado River. When water in a river is diverted upstream, protests can be heard if communities downstream feel that insufficient water is being left to provide for their water requirements and to protect river fisheries. Eventually, this nation of extravagant water users may have to place limits on the use of this undervalued and misunderstood resource.

As urban sprawl has encroached on watershed areas, the pollution of reservoirs has begun to loom as a problem. As villages have grown into towns, more sewage has begun to be washed into reservoirs and their tributaries. Runoff of fertilizer, pesticides, animal wastes, and petroleum products from farms, yards, and the increasing number of roads and other paved areas produces additional contamination. Consequently, cities and states have placed stringent limits on the property-use rights of landowners in watersheds, frequently without compensation to the property owners. Because these limits on property use have in many cases caused land values in watershed areas to decrease, there is increasing conflict between watershed communities and the towns and cities that obtain their water supply from those watersheds.

## ENERGY

In primitive societies people require fuel only to produce energy for heating and cooking, but in more developed societies people also need fuel to produce energy for transportation, electric power generation, and industry. As Third World countries become more industrially developed and as increasing numbers of people in these nations become more affluent, the amount of energy they use will undoubtedly also rise, because energy use increases not only with population growth but also with technological development and the improvement of living standards.

Americans account for only 4.5 percent of the world's population, yet we use more than 20 percent of the energy consumed in the world each year.[2] Although the rate of population growth in the United States has declined in recent years, the per capita use of energy has continued to rise.

In the United States, most of the energy employed for transportation as well as a large portion of that used for heating and for generating electricity comes from oil. During the past few decades, as much of the easily extractable oil in the oil fields in the lower forty-eight states has been exhausted, the country has become increasingly dependent on oil located in formerly undisturbed wilderness areas (Alaska, for example) and on oil imported from places where it can still be extracted cheaply. Furthermore, since more than half the oil used in the United States is transported great distances in huge ships, the risk of environmental damage from oil spills as a result of shipping accidents has increased.

In the 1950s and 1960s, nuclear power seemed to offer great hope as a source of cheap, clean energy. Aside from the volumes of hot water released into adjacent rivers, bays, and oceans, there seemed to be little negative environmental impact. But as the dangers of nuclear accidents became more obvious and as the risks of storing and managing relatively large quantities of radioactive wastes over long periods of time became increasingly evident, atomic power slowly lost its luster.

Both water and wind have been used as power sources in America since the colonial period. Today, about 5 percent of the power consumed in the United States comes from major hydroelectric facilities such as the Tennessee Valley Authority and the Glen Canyon Dam.[3] However, because many people are beginning to think that the environmental damage caused by such large dams outweighs the benefits, few new large hydroelectric plants are apt to be built and some older ones in need of costly repairs may eventually be dismantled. In recent years, serious consideration has been given to removing dams from small rivers across the country, and more than a dozen dams have already been breached.

Energy from the sun and wind, which does not create byproduct pollutants (save in the manufacture of the apparatus for collecting it), may yet have a future in the United States, but this will probably require stronger governmental support. At present, less than 1 percent of the energy consumed in this country comes from solar collectors and windmills.

Although environmentalists have called for better public transportation and vehicles that are more fuel efficient, urban sprawl mandates ever greater dependence on the automobile. Over the years, cars have become more fuel efficient, but the ever-increasing number of cars on America's, and the world's, roads has counteracted the gains in air pollution control achieved through increased fuel efficiency.

## AIR QUALITY, CLIMATE CHANGE, AND ATMOSPHERIC ISSUES

When fossil fuels are burned to produce energy, carbon dioxide, sulfur gases, and nitrogen gases are released into the atmosphere. Some of these waste gases then combine with the normal components of air, such as oxygen, and form acids. The presence of these acids in the air results in acid rain or smog. Acid rain causes widespread damage to forests and life in lakes and rivers; smog is responsible for or contributes to impaired respiratory function, especially in the very young and the very old. Although early efforts to control the emission of polluting gases into the atmosphere were relatively successful, as the concentration of pollutants has been reduced, the cost of eliminating the remaining pollutants has increased. The increasing cost of eliminating smaller and smaller quantities of air pollutants compared with the decreasing health benefits of their elimination has been the focus of rancorous debate.

Many scientists think that the increasing amount of carbon dioxide in the atmosphere (it has increased by 30 percent since the 1850s) is causing a warming of the earth's atmosphere. Other people, including some scientists, believe that the warming trend that has been recorded in recent decades is merely the result of normal temperature variations. No matter what the cause, a rise in temperature and the accompanying change in the distribution of rain will have a major effect on agriculture and the way we live. As the earth warms, glaciers will melt and sea levels will rise. Although many of us find it hard to worry about tides a few more inches above sea level, people in coastal communities recognize the devastating potential of this seemingly minor change. Indeed, island nations as well as countries with large areas of lowlands, such as Bangladesh, may lose substantial portions of their national territory.

Environmentalists think that it is unconscionable for nations to fail to act to control carbon emissions even if their relation to global warming is unproved. Industrialists, on the other hand, believe it is foolish to expend vast sums of money and undermine economic prosperity in order to address the remote possibility of environmental danger. In the international arena, where more and more nations are questioning the rights of other countries to pursue policies that may have a deleterious effect on the environment of their nation, the control of carbon dioxide emissions is a controversial issue.

## WASTE PRODUCTION AND DISPOSAL

Since organic wastes can be digested by bacteria both on land and in water, small amounts of wastes from the life processes of various organisms, including humans, are generally recycled in the environment without event. However, because feces and other bodily wastes are potential carriers of diseases, when these wastes accumulate in large quantities their disposal poses a problem. As the nation's population has grown, and as cities and towns have expanded, the need to find safe and adequate ways to dispose of ever-increasing amounts of both organic and inorganic wastes has presented an increasingly formidable challenge.

Most cities and towns in the United States dispose of household waste water by releasing it into streams and rivers, and until relatively recently much of this sewage was released without any kind of treatment. Huge quantities of wastes from farm animals also wash into rivers and streams. In areas where human or animal populations are large and untreated sewage is allowed to flow into waterways, the oxygen levels in these waters are reduced as a result of bacterial oxidation, often to the point where fish and other aquatic organisms suffocate and die.

Today, waste management to reduce or eliminate the harmful effects of sewage release in urban and suburban areas is required by federal law. However, many cities still have sewer systems that merge household wastes with rainwater runoff from streets and buildings. Although the systems may be able to manage all the household wastes, they frequently cannot manage both household sewage and concurrent runoff. The result is that whenever there is heavy precipitation, such systems are overwhelmed and untreated waste flows into waterways. To resolve this problem, cities are faced with the prospect of immense capital expenditures to build structures that will temporarily store millions of gallons of runoff mixed with sewage until it can be processed at a waste treatment facility.

In the process of producing and using the materials, machines, and foods that we employ in our everyday lives, industrial plants and individuals produce wastes ranging from vast quantities of hot water (which flow from cooling systems), to gases (including carbon dioxide, nitrogen dioxide, and sulfur dioxide), to various solid wastes (including minute quantities of such toxic elements as lead and mercury). These wastes may be carried in the water we drink, in the air we breathe, or in the flesh of animals we eat. Some of these wastes are serious pollutants that can have harmful effects at a variety of levels, from the cellular level to the ecosystem level. Although some release of pollutants is the result of ignorance (such as the release of lead from leaded gasoline, which was widely used until the 1970s) or accidents (such as chemical and oil spills), some is the result of the illegal dumping of known toxic substances. Because of the expense and difficulty of proper disposal of toxic wastes, toxic substances often have been deliberately disposed of illegally. However, the cost of dealing with contamination from improperly discarded toxic and radioactive substances is enormous compared to that for properly handled toxic waste.

Efforts to remedy pollution resulting from what are designated as "hazardous waste sites" have not been entirely successful despite the passage of laws and huge expenditures of funds. Conflicts have arisen between those who claim that a particular hazardous waste site is the cause of a high incidence of an illness (such as a particular type of cancer) in an area and those who dispute the claim and view the costly cleanup of the waste site as unnecessary.

As the population increases, more garbage is produced and more sites are needed to dispose of it. Since no one wants these dump sites and incinerators in his or her backyard, a disproportionate number of these waste disposal sites have ended up in the backyards of the people with the least political clout—the poor and members of racial and ethnic minority groups. This obvious injustice has provoked a discussion of the issue of environmental equity.

In the 1990s, many towns and cities attempted to reduce the amount of waste moving onto disposal sites by recycling bottles, cans and other metal objects, paper, and some plastics. Although some wastes can be recycled, the recovery process itself uses energy, and not all the material can be recovered for reuse.

## TOXIC CHEMICALS

Waste disposal sites and factories are not the only major sources of toxic chemical pollution. Indeed, the world has always contained toxic substances. Poisonous plants and animals as well as toxic minerals such as lead and mercury have always had the potential to harm living things. However, natural processes generally do not concentrate these naturally occurring toxic materials. In contrast, many industrial processes and products developed since the late eighteenth century do just that.

Until recently, the concentrations of lead in gasoline and house paint and of mercury waste from industrial processes were at levels high enough to cause a variety of illnesses in humans. Other toxic chemical problems have arisen from the use of substances not known to occur in nature. Unlike most natural toxins, many complex chemical compounds developed during the past century are not readily biodegradable (i.e., they do not deteriorate naturally at all, or they deteriorate very slowly). Consequently, several toxic man-made substances such as DDT, an insecticide, and polychlorinated biphenyls (PCBs), chemicals used to moderate heat in electric transformers, have accumulated in the environment, causing serious problems for wildlife.

## FORESTS, WILDERNESS, AND WILDLIFE

When the first European settlers arrived, most of North America was covered by forests. A vast deciduous forest stretched from the East Coast to the Mississippi, and another massive forest extended west from the Rocky Mountains to the Pacific Coast. Only the arid Southwest, the desert lands, and the Great Plains were without forests. Although Native Americans had made some changes to the original growth forests by burning and cutting the forest in places and clearing the understory, much of the original forest was still intact. Today, most of the original forest is gone. As the population of the United States increased, large segments of the original forests were cut down to make way for farms and to provide wood for fuel and construction material.

In much of the eastern part of the United States, forests are returning for several reasons. Wood is no longer used for fuel on a substantial level; wood for construction is increasingly coming from cultivated plots; and the number of farms and farmers has decreased, allowing farmland to lie fallow and return to forest. Although forested lands have increased in recent years, the nature of the forest has changed. Where once there were thousands of acres of uninterrupted forest, now we see thousands of acres of forest segmented by roads and fields. As a result, many woodland species are no longer protected from animals that inhabit the edges of forests.

In several sections of the country, growing trees to obtain wood for use in construction or in the production of paper has become much like farming corn or wheat. Landowners, who are often large corporations, plant thousands of acres with one species of tree to be harvested all at once. This type of tree farming and harvesting, known as monoculture and clear-cutting, generates a forest that lacks the diversity of forests native to the area. The complete removal of the forest during clear-cutting not only leaves those few animals that are able to adapt to the monoculture without shelter, but also leaves the former forest floor vulnerable to soil erosion and degradation. If all the trees are removed from an area, the land is exposed to direct sunlight, and the heat from the sun—which would have been absorbed by the trees if the land had remained forested—is absorbed by the soil. When rain falls on the exposed land, it is heated by the warm soil and, because there are no tree roots to hold the soil, the warm runoff from the rain carries the topsoil into adjacent rivers and streams. Where the runoff from the forest used to provide clear, cool water for the river or stream, the clear-cut area now supplies warm water with soil particles suspended in it. Since warm water is incapable of carrying as much oxygen as cool water, the amount of oxygen available to the aquatic creatures in the stream is reduced, and as a result, much of the aquatic life of the region will die.

In other parts of the country, corporations harvest trees from naturally growing forests that are on lands owned primarily by the federal government. The corporations tend to harvest the lumber in the most efficient (cheapest) way—by clear-cutting. Opposition to using federal forests in this manner comes from people who are concerned about wilderness. They cite the giveaway of valuable timber, the loss of habitat for resident species, and the impact of clear-cutting not only on the forests themselves but also on the streams that run through them. Along the Pacific Coast, conservationists, fishermen, and lumbermen quarrel over the effects of disturbing mature forests where spotted owls nest in ancient trees and salmon are dependent on clear, cool streams for spawning.

## FISHERIES AND AQUATIC LIFE

This nation, with a heritage of rich fishery resources, is heading into the twenty-first century with problems in all of its aquatic quarters. Shellfish environments have been polluted and depleted by sewage from nearby cities; vast beds of oysters and clams are off limits to harvesters because of the potential presence of pathogens. The great runs of spawning fish, which return from the ocean along both coasts of the United States, have been reduced, and in some places eliminated, by pollution and the damming of rivers.

In the East, few rivers with salmon runs remain free of critically placed dams. Many eastern rivers are only now beginning to recover from severe industrial pollution that lasted for nearly a hundred years, from the 1880s to the 1980s. In other places, residual toxins in river bottoms continue to circulate in the water column and render the fish inedible. The Hudson River, which serves as a major spawning ground and nursery for huge populations of striped bass, was a thriving commercial fishing center until documentation of carcinogenic PCBs in the tissues of fish caught in the Hudson ended commercial fishing in the river.

In the oceans, long viewed as inexhaustible sources of fish and other seafood, the problem seems even worse. As fishing technology has "improved," the fish supply has come under tremendous pressure. New fish-finding techniques and harvesting methods have interrupted the life cycles of the fish and pushed the populations below the level of sustainability in ocean areas where generations of fishermen once supported themselves and their families by fishing from small, unsophisticated boats. In the global marketplace, astronomical prices are offered for large predatory fish, such as bluefin tuna and swordfish, driving these now hard-to-find fish to the brink of extinction. With fishermen harvesting fewer and smaller fish, governments are struggling both to protect the fish and to keep the fishermen employed. Conflicts over how best to manage regional fisheries pit environmentalists who insist on catch limits to allow the replenishment of fish stocks against fishermen who need to earn a living.

## THE LOSS OF BIODIVERSITY

The earth is populated with billions of different species of living things, ranging from an enormous variety of bacteria to hundreds of species of orchids to a multitude of different kinds of primates, from gorillas to humans. Where and how each species lives is a product of geologic and climatic factors as well as the presence of other living things in its habitat.

These biologically diverse organisms are dependent on one another for the food they eat and the well-being of the ecosystems in which they live. The interdependence of living things results in there being a maximum and a minimum number of each species that can exist within a particular ecosystem without upsetting what has been called the "food chain," the "web of life," and the "balance of nature."

Changes in the kinds and number of members of species that live in particular habitats occur naturally. Climatic factors, such as fluctuations in rainfall and temperature, and geological upheavals, such as volcanoes and earthquakes, can cause the decline, extinction, or movement of species. The introduction of invasive foreign species (which can be brought as seeds in the feathers, fur, or feces of migrating species or result from animal migrations caused by geological upheavals or climatic changes) may also cause species extinction and result in disruptions in the food chain.

Human civilization, by making major alterations in the natural environment and by replacing a varied environment with a much more uniform environment, has been responsible for the extinction of numerous species. Towns and cities replace wild habitat; farmers, seeking to maximize food supply or income, grow only edible or cash crops and have eliminated all other plants from the land; the timber industry clear-cuts whole forests of diverse species of trees and then plants new forests containing only a single species. As the habitat becomes more homogeneous, the number and variety of species that can be sustained by it grows smaller. Many people believe that as human civilization has developed and expanded, the rate of species extinction has accelerated, and they are alarmed that the current rate of species extinction is unparalleled in the recent history of the earth.

There are several reasons for concern about species extinction and the resultant loss of biodiversity. First, as diversity is lost, the resilience of the remaining species in the entire area is reduced, and as a result, their population decreases. Then, if some kind of blight or disease occurs that affects a particular species, the loss of that species is likely to have a much more devastating effect on the whole region. Second, the loss of biodiversity results in the loss within species of individuals that carry varied genetic material. The wild strains of many domesticated plants retain genetic material that has been eliminated from the current domesticated varieties but that may be of value in rejuvenating domesticated strains in the future (for example, if the domesticated strain became vulnerable to some kind of blight or if there was a need to develop a more heat tolerant strain). As the world becomes more and more dependent on a smaller and smaller variety of plants and animals for food, timber, and natural fibers, there is increasing concern about the loss of genetic variation. Third, environmentalists warn that we are losing species that may be of great value to humans, and that many of these plants or animals may not even have been discovered.

Because biodiversity reaches its maximum in rain forests, when rain forests are destroyed, incredibly diverse and potentially valuable communities are lost. Opposition to rain forest destruction is particularly strong among environmentalists, but for loggers, miners, and farmers who look to the rain forest for their daily sustenance, the long-term effects of rain forest destruction are not a major concern.

# Part I

## Foundations of American Environmental Thought and Action

Some thirty thousand years ago, when Asia and North America were still connected by a land bridge in the area of the Bering Strait, a few Asian hunters and gatherers crossed this narrow stretch of land between what are now Russia and Alaska and became the first humans to set foot in the Western Hemisphere.[1] Like other people who migrated to the Americas during the ensuing millennia, these prehistoric men and women probably came in pursuit of food or more hospitable surroundings. Possibly they were tracking a wild beast, searching for a special plant, or fleeing from competing bands of humans or predatory animals when they ventured into this vast, unpeopled land. As word of game and edible plants in the newly discovered region filtered back into Asia, additional groups of migrants made their way to the Western Hemisphere. Eventually some of these people drifted south along the western coast of the great land mass.

Then, about twenty-two thousand years ago, the climate began to change. Slowly, glacial ice spread down from the north, covering the land bridge and trapping these early settlers on the eastern shores of the Pacific. As the ice spread over the whole northern part of the continent, humans, animals, and plants moved south into Central and South America.

When the climate changed again about thirteen thousand years ago and the ice receded, some of the descendants of the early settlers began moving north, bringing with them, from what are now Peru and Mexico, a plant called maize and, eventually, the agricultural technology to cultivate it. When the ice receded even farther, new waves of Asian immigrants once again began crossing the Bering land bridge and spreading southward and eastward across North America.

By the time Columbus arrived in the "New World" in 1492, the indigenous population had grown to an estimated 7 million people, living throughout North and South America and the Caribbean. The approximately 1 million inhabitants of North America occupied every part of the continent, from the ice-bound north to the dry southwest and the luxuriant lands of the Gulf Coast, from the forests of the northwest to those of the northeast, from the grassy central plains down to the rich, watery lands of the southeast and the arid lands of the southwest. In the Great Plains region, it is estimated, there were about 225,000 people, and west of the Rocky Mountains perhaps 350,000 people.[2] Most tribes throughout the continent practiced some form of farming and relied on hunting, fishing, and foraging to satisfy the balance of their material and dietary needs. They had no iron for tools, no horses or wheels for transportation, and in the northern regions (in what would become the United States and Canada) no written language or knowledge of higher mathematics.

## THE CONTACT BETWEEN EUROPE AND AMERICA

Across the Atlantic Ocean, in Europe, a great renaissance had begun in about the fourteenth century. A surge of economic growth and an intellectual blossoming during this period resulted in advances in science and map making. The increased intellectual openness, as well as the spread of knowledge that followed the invention of printing with movable type in about 1450, made possible the technological advances, including advances in maritime technology, that led to an expansion of maritime trade and the great voyages of discovery of the late fifteenth and sixteenth centuries.

Spanish explorers established bases in the Caribbean region early in the sixteenth century. By the middle of the century, they had sailed up and down the Atlantic and parts of the Pacific coasts of North America, had made forays into much of the territory that eventually became the southern part of the United States, and had set up a permanent colony at St. Augustine, Florida. They had also tried, unsuccessfully, to establish permanent settlements along the Atlantic coast as far north as what is now South Carolina. Other explorers, sailing under the flags of various European states, including England and France, had also explored the Atlantic seaboard. Both the French and the English had attempted to establish colonies along the Atlantic shore during the sixteenth century, but resistance from the native inhabitants, disease, and lack of food proved to be insurmountable obstacles to the survival of these colonies. It was not until 1609 that the English gained a permanent foothold on the western shores of the Atlantic with the establishment of a settlement at Jamestown, Virginia.

During the century that followed the successful colonization of Jamestown, the landscape of North America was transformed by an influx of permanent settlers from western Europe. Colonies were established all along the Atlantic seaboard, and traders, trappers, and missionaries traveled deep into the interior of the continent. By 1700 the territory that would eventually form the thirteen rebellious British colonies had a population approaching 300,000.

## THE CLASH OF CULTURES

The vast majority of the fifteenth- to eighteenth-century explorers and colonists of North America—whether they came from Spain, England, France, the Netherlands, or elsewhere in Europe—brought with them very similar attitudes about the relationship between humans and the natural world. Predominant among these was a belief that humanity is at the center of creation, that people have a right to use the resources of the land for human benefit, and that it was their divine duty to subdue the land they had discovered and the non-Christians who occupied it. For the most part, the natural world was seen as either a beneficent garden with riches created for the exclusive use of humans or as a savage, evil wilderness to be conquered and tamed. Those parts of the world that had been occupied and cultivated were viewed as potential gardens, while the uninhabited wilderness was looked on with fear and mistrust.

The way the European explorers and colonists regarded the world and the roles they envisioned for themselves were shaped by the Bible, the classics, and tradition. The Bible, a work familiar to all of the early explorers and fundamental to the education of the colonists, was a primary sourcebook [see Document 1]. Over the centuries, a wide range of other writings had also greatly influenced the mind-set of the Europeans, including classics such as the *Eclogues* of Virgil [see Document 2] and the scientific and philosophical writings of Francis Bacon [see Document 8] and Isaac Newton. It is doubtful that many of the colonists would have been familiar with the writings of St. Francis of Assisi,[3] whose vision of the brotherhood of all creation closely paralleled the views of many North American Indians but was an anomaly in European thought until recently.

The Europeans considered the Native Americans as either merely another resource created for the use of civilized humans (Europeans) or as savage, uncivilized creatures to be Christianized. However, despite the European interlopers' condescending attitude toward the Indians, the explorers and colonists were clearly dependent on the natives for knowledge about the weather, plants, and animals in the New World [see Document 7].

The newcomers were intent on claiming for themselves—both as representatives of their sovereigns and as individuals—as much of the land and its wealth as they could lay their hands on [see Documents 3–5]. However, the Native Americans were steadfast in their desire to continue to farm, hunt, and fish on the lands and in the rivers where they had done so traditionally. As the growing white population appropriated more and more of America's land and wildlife, the native tribes found themselves with access to fewer of the resources essential to their well-being and continued existence, and the conflict between the Indians and the Europeans intensified.

The Indians, who had no concept of private property rights and did not consider land and water to be transferable assets, probably assumed that they would continue to have access to the resources of a place as long as the land was unoccupied, even if it had been sold [see Document 32]. They viewed themselves as part of a great whole, as just one of the many inhabitants of the earth, along with the birds and the four-legged animals, and they saw the wilderness as an integral part of the natural order.

Europeans and colonists, on the other hand, had little appreciation of or respect for the wilderness. This was true even of the great naturalists of the eighteenth century, both European and American born. Although they may have been impressed with the variety of plants found in the wilderness, they considered the wilderness's major value to be its potential as a source of plants that could be cultivated on farms and in gardens or as land that could be transformed into farms and gardens [see Documents 15, 17, and 19].

In spite of their belief that the resources of the land were a God-given gift to humans and their sense of being separate from the natural world, the colonists were forced to recognize, within just a few decades of their arrival in America, that, without planning, some of the resources of their new land might soon become scarce. By the mid-seventeenth century the colonists had begun to institute laws regulating the use of timber, fish, and game animals [see Documents 9, 10, and 13], and even to limit pollution [see Document 12]. In the English colonies it was standard practice to set aside common land for grazing and timber [see Document 13], but the commons had to be regulated to ensure that no individuals would use more than their allotted share of the common resources.

## THE CLASH OF ECOSYSTEMS

The arrival of the Europeans in the Americas produced a clash not only of two very distinct cultures but also of two separate ecosystems. In the holds of their ships, the Europeans carried horses, pigs, cattle, sheep, and chickens, as well as crop seeds and fruit trees [see Document 5]. They also brought, on their clothes and boots, the seeds of blue grass, dandelions, and daisies and, in their bodies, the microorganisms that cause smallpox and measles. When they returned to Europe they took with them beans, maize, potatoes, and tomatoes. In America, the horse (first brought to the "New World" by the Spaniards in 1493) gave the Europeans a huge military advantage and then transformed the lives of the Indians. But it was European diseases as much as the Europeans' military strength that caused the decimation of the Native American population [see Document 6]. Some scholars have estimated that as much as one-half of the precontact population of the Americas had died of smallpox within a few decades of Columbus's landing. In Europe, on the other hand, the introduction of the potato changed the continent's agricultural base and helped to fuel population growth.[4]

The increase in Europe's population created a need for more agricultural and grazing lands, while the expanding maritime industry required increasing amounts of timber. As a consequence, the great forests that at one time covered much of England and the European continent slowly began to fall to the ax. Simultaneously, Europe's industrial growth produced a need for increasing quantities of raw materials for its factories and mills. The Western Hemisphere provided a haven for Europe's burgeoning—and increasingly urban—population, a source of raw materials, and a market for its industrial products.

## FORGING A NEW NATION

While the fundamental cause of the American Revolution was resistance to taxation without representation, the revolutionary spirit was fostered by colonial resentment of England's dumping of manufactured goods and unwanted people (including criminals) on American land, imposed limits on manufacturing, and a sense that the exploitation of America was being carried out for the benefit of England with little regard for the colonies' inhabitants.

In spite of their differences with England, the rebellious colonists looked to England, as well as to France, for legal precedents and philosophical ideology as they prepared to launch a new nation. And when they required a philosophical and legal basis for depicting the human relationship with the land, they turned to the writings of Europeans such as Thomas Hobbes [see Document 11], René Descartes, Jean-Jacques Rousseau, John Locke [see Document 14], and William Blackstone [see Document 18].

---

## DOCUMENT 1: Biblical Views of Nature and Humanity

Many believe that the foundation for human abuse of the environment lies in the story of creation at the beginning of Genesis, where humans are directed to subdue the earth and exert dominion over it. Although there has been extended debate about the meaning of "dominion," there is little doubt about the interpretation given to the term by the newcomers to North America. The new Americans were to become adept at "subduing" the earth, but good at "replenishing" it only with increasing numbers of people.

However, the biblical story of Noah, which describes Noah's management of the flood according to God's instructions, suggests another role for humanity in relation to the natural world. The sort of dominion Noah exercised over the creatures that he took onto the ark indicates respect for the inherent value of nature and a sense of a reciprocal relationship between living things. If this act of saving all living things can be considered "dominion," it surely contrasts with the interpretation given the biblical injunction by the colonists and later Americans.

In Isaiah, as well as elsewhere in the Bible, the contrast between the garden and the wilderness is vividly described. This dichotomy, although it may accurately reflect the environment in which the biblical prophet lived, seems inappropriate when applied to eastern North America. Nevertheless, the language of this dichotomy was carried across the Atlantic and embedded in the thinking of the colonists.

### A. Genesis: The Story of Creation

So God created man in his own image, in the image of God created he him; male and female created he them.

And God blessed them, and God said unto them, Be fruitful, and multiply, and replenish the earth, and subdue it; and have dominion over the fish of the sea, and over the fowl of the air, and over every living thing that moveth upon the earth.

And God said, Behold, I have given you every herb bearing seed, which is upon the face of all the earth, and every tree, in the which is the fruit of a tree yielding seed; to you it shall be for meat.

## B. Genesis: The Story of Noah and the Flood

And behold, I, even I do bring a flood of waters upon the earth, to destroy all flesh, wherein is the breath of life, from under heaven: *and* every thing that *is* in the earth shall die.

But with thee will I establish my covenant: and thou shalt come into the ark, thou, and thy sons, and thy wife, and thy sons wives with thee.

And of every living thing of all flesh, two of every *sort* shalt thou bring into the ark, to keep them alive with thee; they shall be male and female.

Of fowls after their kind and of cattle after their kind, every creeping thing of the earth after his kind: two of every *sort* shall come unto thee, to keep *them* alive.

. . .

And God remembered Noah, and every living thing, and all the cattle that *was* with him in the ark: and God made a wind to pass over the earth, and the waters asswaged.

. . .

And God spake unto Noah, saying,

Go forth of the ark, thou, and thy wife, and thy sons, and thy sons wives with thee.

Bring forth with thee every living thing that *is* with thee, of all flesh, *both* of fowl, and of cattle, and of every creeping thing that creepeth upon the earth; that they may breed abundantly in the earth, and be fruitful, and multiply upon the earth.

## C. Isaiah, c. 725 B.C.E.

They shall lament for the teats, for the pleasant fields, for the fruitful vine.

Upon the land of my people shall come up thorns and briers, yea, upon all the houses of joy in the joyous city:

Because the palaces shall be forsaken, the multitude of the city shall be left, the forts and towers shall be dens for ever, a joy of wild asses, a pasture of flocks.

Until the spirit be poured upon us from on high, and the wilderness be a fruitful field, and the fruitful field be counted for a forest.

Then judgment shall dwell in the wilderness, and righteousness remain in the fruitful field.

. . .

[T]he Lord shall comfort Zion: he will comfort all her waste places, and he will make her wilderness like Eden, and her desert like the garden of the Lord; joy and gladness shall be found therein, thanksgiving, and the voice of melody.

*Source*: Holy Bible, Vol. 1 (Philadelphia: R. Aitken, 1782), Genesis 1:27–29, 6:17–20, 8: 1, 15–17; Isaiah 32: 12–16, 51: 3.

---

## DOCUMENT 2: Virgil's Pastoral View of Nature (c. 50 B.C.E.)

From the age of the Greeks and Romans through the Renaissance, nature was considered to be at its most sublime when molded by the human hand and turned into farmland, gardens, or parks—land that provided both serenity and sustenance. This domesticated nature was celebrated in pastoral poetry, such as that of the Roman poet Virgil, whose *Pastorals*, or *Eclogues* as they are frequently called, depict the domesticated countryside as a place of joy for those who live there and a place yearned for by those who do not. In this extract, Meliboeus, who has been turned out of his land in a redistribution of property by the Roman emperor Augustus, is speaking to Tityrus, who has been permitted to stay.

The celebration of nature in poetry and other writings is probably as ancient as literature itself. Such writing usually expresses a timeless human sense of wonderment about the beauty or awesomeness of nature; it also reflects a sense of the human relationship with the natural world at a particular time and place. The pastoral ideal may have colored the dreams of some early Americans, but it was inappropriate for the American landscape. Toward the end of the eighteenth century, American writers and painters began to reject this pastoral ideal and turn to untamed nature for inspiration [see Document 22], and by the mid-eighteenth century the romantic appreciation of wilderness had become a dominant influence in American art and literature [see Document 34].

> Beneath the shade which beechen boughs diffuse,
> You, Tityrus, entertain your silvan muse.
> Round the wide world in banishment we roam,
> Forc'd from our pleasing fields and native home;
> While, stretch'd at ease, you sing your happy loves,
> And Amaryllis fills the shady groves.

* * *

> O fortunate old man! Whose farm remains—
> For you sufficient—and requites your pains;
> Though rushes overspread the neighbouring plains,
> Though here the marshy grounds approach your fields,

And there the soil a stony harvest yields.
Your teeming ewes shall no strange meadows try,
Nor fear a rot from tainted company.
Behold! yon bordering fence of sallow trees
Is fraught with flowers, the flowers are fraught with bees:
The busy bees, with a soft murmuring strain,
Invite to gentle sleep the labouring swain.
While, from neighbouring rock, with rural songs,
The pruner's voice the pleasing dream prolongs,
Stock-doves and turtles tell their amorous pain,
And, from the lofty elms, of love complain.

*Source*: Virgil, "Pastoral I," in *Dryden's Version of Virgil's Pastorals and Georgics; and the First Volume of Aeneis*; Vol. 10 of *The Works of the Greek and Roman Poets, Translated into English* (London: Suttaby, Evance and Fox, 1813), pp. 37, 39–40.

## DOCUMENT 3: Christopher Columbus Inventories the New World's Natural Resources (1493)

In this letter to Luis de Santangel, comptroller of the treasury of King Ferdinand and Queen Isabella of Spain, Columbus describes the riches of the Caribbean islands he has discovered and extols the opportunities for the exploitation of their abundant resources. To Columbus, Jean Ribaut [see Document 4], and other early explorers, the "New World's" resources appeared to be limitless, and for nearly four centuries most Americans believed that this was indeed the case.

I discovered many islands inhabited by people without number, and of which I took possession for their Highnesses by proclamation with the royal banner displayed, no one offering any contradiction. . . . All these countries are of surpassing excellence, and in particular *Juana*, which contains abundance of fine harbours, excelling any in Christendom, as also many large and beautiful rivers. . . . [The mountains] are accessible in every part, and covered with a vast variety of lofty trees, which it appears to me, never lose their foliage, as we found them fair and verdant as in May in Spain. Some were covered with blossoms, some with fruit, and others in different stages, according to their nature. The nightingale and a thousand other sorts of birds were singing in the month of November wherever I went. There are palm-trees in these countries, of six or eight sorts, which are surprising to see, on account of their diversity from ours, but indeed, this is the case with respect to the other trees, as well as the fruits and weeds. Beautiful forests of pines are likewise found, and fields of vast extent. Here is also honey, and fruits of a thousand sorts, and birds of every variety. The lands contain mines of metals, and inhabitants without number. The island of Espanola is pre-eminent in beauty and excellence, offering to the sight the most enchanting view of mountains, plains, rich fields for cultivation, and pastures for flocks of all sorts, with situations for towns and settlements. ... [T]he preference [among the islands discovered] must be given to Espanola, on account of the mines of gold which it possesses, and the facilities it offers for trade with the continent, and countries this side. . . . At present there are within reach, spices and cotton to as great an amount as they can desire, aloe in as great abundance, and equal store of mastick. . . . To these may be added slaves, as numerous as may be wished for. Besides I have as I think, discovered rhubarb and cinnamon, and expect countless other things of value will be found.

*Source*: Christopher Columbus to Luis de Santangel, in *Personal Narrative of the First Voyage of Columbus to America* (Boston: T. B. Wait, 1827), pp. 253, 255–56, 260, 263.

---

## DOCUMENT 4: Jean Ribaut Discovers the Natural Abundance of Terra Florida (1563)

Like other explorers and colonists both before and after him, Jean Ribaut, a captain in the French navy, was impressed by the riches of the "New World" and depicts it as a paradise abounding in honey and venison. He sought to claim and exploit as much territory as possible for the nation under whose flag he sailed, and he established a colony at what is now Port Royal, South Carolina.

We entered and viewed the countrey thereabouts, which is the fairest, fruitfullest, & pleasantest of all the worlde, abounding in honye, venison, wylde foule, forests, woods of all sorts, Palme trees, Cypres & Cedres, Bayes the highest & greatest, with also the fairest vines in all the world, with Grapes according, which without naturalle arte & without mans helpe or trimming wil growe to toppes of Oks, & other trees that be of wonderfull greatnesse and heyght. And the syght of the fayre Medowes is a pleasure not able to be expressed with tongue: full of Hernes, Curlues, Bitters, Mallardes, Egreyths, Wodkockes, & all other kynde of small byrdes: With Hartes, Hyndes, Buckes, wylde Swine, & all other kyndes wylde brathes, as we perceyved well bothe by theyr footing there, & also afterwardes in other places, by theyr crye & roryng in the nyght.

Also there be Connies & Hares: Silke wormes in mervelous number, a great dell fairer & better then be our silk wormes. To be short, it is a thing unspeakable to consider the thinges that be seene there, & shalbe found more & more in this incomparable lande, whiche never yet broken with ploughe was [but] bryngs forthe all things according to its first nature, wherewith the eternal God endowed it. About theyr houses they [the Indians] labour & tyll the grounde, sowyng theyr fields with a grayne called Mahis, whereof they make theyr meale: & in theyr Gardens they plant beanes, gourdes, cocumbers, citrons, peason & many other fruites & rootes unknown unto us. Their spades: mattocks made of wood, so well & fitly as is possible: which they make wyth certayn stones, oyster shells & muscles, wherewith also they make theyr bowes & small launces: & cutte & polyshe all sortes of Wood, that they imploye aboute theyr buildings, & necessarie use: There groweth also manye Walnut tress, Hasell tress, Cheritrees, very fayre and great.

*Source*: Jean Ribaut, *The Whole and True Discovereye of Terra Florida*, trans. Thomas Hacket (London: Rouland Hall, 1563), unfolioed.

---

## DOCUMENT 5: Baltasar de Obregon's *Account of the Riches of New Mexico* (1584)

In the latter part of the sixteenth century, Spaniards who were based in Mexico made several expeditions to the lands of the Pueblo Indians. The Chamuscado-Rodriguez expedition of 1582, mentioned in this document, was one of the most important of these ventures. Although the primary object of these forays was to find gold, silver, and copper and to record the locations of mines, the expedition reports also detailed how the Indians interacted with their environment and evaluated the potential for introducing European crops and domesticated animals.

The natives gather quantities of corn, beans, calabashes, cotton, and *piciete*, a very useful herb. They make large numbers of blankets, both heavy and light, beautifully woven and dyed with various bright colors. They possess numerous turkeys. They utilize the feathers, interweaving them in heavy cotton blankets. They have quantities of salines of rich salt. There are salt deposits that extend over five leagues. They have large numbers of mats made of rushes and reeds, and large and small baskets. They possess good crockery, both heavy and fine, brilliantly decorated with admirable colors. They grow Castile flax without cultivation. It flourishes naturally at Cieneguilla and the Valle de los Valientes. Consequently they make Castile cloth.

... Thirty or forty leagues away are numerous cattle which they utilize; the meat for food and the hides for many purposes like the hides of
the cattle of Spain. They use their wool for clothing, the fat for candles
and other things. The hides are good for making shoes and weapons
when very well tanned. These provinces and towns mentioned have a
fine climate, numerous plains, valleys, mountains, rivers, streams, lakes,
springs and riverbanks, suitable for the cultivation of any kind of grain
from Spain and for raising all sorts of cattle.

... There are many sierras on its confines where I saw and examined
rich metals when I was on the expedition with General Francisco de
Ibarra. In the ridges of these mountains and near the settlements are the
mines discovered and inspected by Francisco Sanchez Chamuscado and
his companions.

*Source*: Baltasar de Obregon, *Obregon's History of 16th Century Explorations in
Western America entitled Chronicle, Commentary, or Relation of the Ancient and Modern Discoveries in New Spain and New Mexico,* trans. and ed. George P. Hammond
and Agapito Rey (Los Angeles: Wetzel, 1928), pp. 300–301.

---

## DOCUMENT 6: Thomas Hariot on the Death of Indians from a Disease Brought from Europe (1588)

In 1585 the Oxford mathematician Thomas Hariot accompanied Sir
Walter Raleigh on his voyage to the "New World" to set up a colony
to be named Virginia in honor of England's virgin queen, Elizabeth.
Hariot's main duties were to observe the customs of the Indians and to
make an accounting of the natural resources in their lands. Like Ribaut
[see Document 4], he was greatly impressed by the abundance of resources. His account details one of the unintended consequences of
the European contact with the Native Americans: the decimation of a
large segment of the indigenous population as a result of "virgin soil
epidemics"—the very rapid spread of pathogens among populations
that had not been previously affected by them. Because Hariot provides
no description of the disease that brought death to the Indians with
whom his expedition came in contact other than to note that it had an
incubation period of a few days, it is impossible to identify the disease.
It could have been smallpox, typhus, or any of a number of other ailments.

One other rare and strange accident, leaving others, will I mention
before I ende, which mooved the whole countrey that either knew or
hearde of us, to have us in wonderfull admiration.

There was no towne where we had any subtile devise practised against us, we leaving it unpunished or not revenged (because wee sought by all meanes possible to win them by gentlenesse) but that within a few dayes after our departure from everie such towne, the people began to die very fast, and many in short space; in some townes about twentie, in some fourtie, in some sixtie, & in one six score, which in trueth was very manie in respect of their numbers. This happened in no place that wee could learne but where wee had bene, where they used some practise against us, and after such time; The disease also so strange, that they neither knew what it was, nor how to cure it; the like by report of the oldest men in the countrey never happened before, time out of minde. A thing specially observed by us as also by the naturall inhabitants themselves.

Insomuch that when some of the inhabitants which were our friends & especially the *Wiroans Wingina* had observed such effects in foure or five townes to follow their wicked practises, they were perswaded that it was the worke of our God through our meanes, and that wee by him might kil and slai whom wee would without weapons and not come neere them.

\* \* \*

[S]ome people could not tel whether to think us gods or men, and the rather because that all the space of their sicknesse, there was no man of ours knowne to die, or that was specially sicke.

*Source*: Thomas Hariot, *Narrative of the First English Plantation* (London: Quaritch, 1893; reprint of 1590 edition), pp. 41, 42.

---

## DOCUMENT 7: William Bradford on Life in the Wilderness (1620, 1621)

Together with a group of fellow Pilgrims, William Bradford sailed from England in search of religious freedom and intent on missionizing the Indians. Their destination was the Massachusetts coast, which had been explored by expeditions led by Captain Bartholomew Gosnold in 1602, and Captain John Smith in 1614. They reached Cape Cod in November 1620, and by the following April many members of the party were ill. Although disdainful of the natives, the Pilgrims would have died of starvation if they had not obtained food from the Pautuxet Indians to make it through the harsh winter and if, in the spring, the Indians had not provided them with seed that was suitable for the Massachusetts soil and climate as well as advice on how to plant it. In this selection, Bradford expresses the fear and distaste for the wilderness that was prevalent among the Pilgrims.

... [A]fter longe beating at sea they [the Pilgrims] fell with that land which is called Cape Cod; the which being made and certainly knowne to be it, they were not a litle joyfull. . . . And the next day they gott into the Cape-harbor wher they ridd in saftie.

\* \* \*

Being thus passed the vast ocean, and a sea of troubles before in their preparation . . . they had now no freinds to wellcome them, nor inns to entertaine or refresh their weatherbeaten bodys, no houses or much less townes to repaire too, to seeke for succoure. It is recorded in scripture as a mercie to the apostle and his shipwraked company, that the barbarians shewed them no smale kindnes in refreshing them, but these savage barbarians, when they mette with them (as after will appeare) were readier to fill their sids full of arrows than otherwise. And for the season it was winter, and they that know the winters of that cuntrie know them to be sharp and violent, and subjecte to cruell and feirce stormes, deangerous to travill to known places, much more to serch an unknown coast. Besides, what could they see but a hidious and desolate wildernes, full of wild beasts and willd men? and what multitudes ther might be of them they knew not. Nether could they, as it were, goe up to the tope of Pisgah, to vew from this wildernes a more goodly cuntrie to feed their hops; for which way soever they turnd their eys (save upward to the heavens) they could have litle solace or content in respecte of any outward objects. For summer being done, all things stand upon them with a wetherbeaten face; and the whole countrie, full of woods and thickets, represented a wild and savage heiw.

\* \* \*

After . . . the shalop being got ready, they set out againe for the better discovery of this place, and the m$^r$ of the ship desired to goe him selfe, so ther went some 30. Men, but found it to be no harbor for ships but only for boats; ther was allso found 2. of their houses covered with matts, and sundrie of their implements in them, but the people were rune away and could not be seen; also ther was found more of their corne, and of their beans of various collours. The corne and beans they brought away, purposing to give them full satisfaction when they should meete with any of them (as about some 6. months afterward they did, to their good contente). And here is to be noted a spetiall providence of God, and a great mercie to this poore people, that hear they gott seed to plant them corne the next year, or els they might have starved, for they had none, nor any liklyhood to get any till the season had beene past (as the sequell did manyfest). Neither is it lickly they had had this, if the first viage had not been made, for the ground was now all covered with snow, and hard frozen. But the Lord is never wanting unto his in their greatest needs; let his holy name have all the praise.

. . .

Anno. 1621.

Afterwards they (as many as were able) began to plant ther corne, in which servise [the Indian] Squanto stood them in great stead, showing them both the maner how to set it, and after how to dress and tend it. . . . Some English seed they sew, as wheat and pease, but it came not to good, eather by the badnes of the seed, or latenes of the season, or both, or some other defecte.

*Source*: William Bradford, *History of Plymouth Plantation 1606–1646*, ed. William T. Davis (New York: Scribner's, 1908), pp. 94–96, 100, 115–16.

---

## DOCUMENT 8: Francis Bacon on Science and Technology (1629)

---

The English statesman, essayist, and philosopher-scientist Francis Bacon established the concept of scientific rationality. He asserted that nature is a machine with no inherent value and proposed that human knowledge should be used to improve on nature and adapt it to human needs. The advancement of technology, Bacon thought, would bring benefit to humans without any negative impacts. Although faith in human innovation would continue to the end of the twentieth century, the negative effects of technology were already evident during the Industrial Revolution, and by the end of World War II, recognition of the destructive potential of technology had forced people to reexamine the relationship of humans, nature, and technology.

[T]he real and legitimate goal of the sciences is the endowment of human life with new inventions and riches.

* * *

[T]he introduction of great inventions appears one of the most distinguished of human actions, and the ancients so considered it; for they assigned divine honors to the authors of inventions, but only heroic honors to those who displayed civil merit (such as the founders of cities and empires, legislators, the deliverers of the country from lasting misfortunes, the quellers of tyrants, and the like). And if anyone rightly compare them, he will find the judgment of antiquity to be correct; for the benefits derived from inventions may extend to mankind in general, but civil benefits to particular spots alone; the latter, moreover, last but for a time, the former forever. Civil reformation seldom is carried on without violence and confusion, while inventions are a blessing and a benefit without injuring or afflicting any.

* * *

Again, let anyone but consider the immense difference between men's lives in the most polished countries of Europe, and in any wild and barbarous region of the new Indies, he will think it so great, that man may be said to be a god unto man, not only on account of mutual aid and benefits, but from their comparative states—the result of the arts, and not of the soil or climate.

*Source*: Francis Bacon, *Novum Organum*, ed. Joseph Devey, in *A Library of Universal Literature, Part I* (New York: P. F. Collier, 1901), pp. 58, 104, 105.

## DOCUMENT 9: Regulating the Herring Run in the Town of Plymouth (1637, 1638, 1639, 1662)

Less than two decades after the Pilgrims' arrival, it became obvious that millers who altered or controlled the flow of water past their mills were interfering with the natural movements of certain fish, notably herring (or alewives), in the stream and that this interference would be detrimental to the entire community because it disrupted the spawning of the fish. As a result, millers were required to let the water flow freely at certain times of the year. The harvesting of the fish was also regulated. The "ware," or weir, was a fence or dam through which the fish could not pass.

The last day of March 1637.
It is concluded upon a Townes meeting that Nicholas [Snow] shall repaire the Hering ware and draw and divide the Hering this yeare and

shall have foure and fourty bushells of Indian corne for his paynes but the Town shall pay him for the boards used about the repaire there.

\* \* \*

At a Townes meeting held the VIth day of February 1638. . . .

It is ordered that . . . the Milner shall observe such order in stoping and loosening of the water as shalbe given by the overseers of the hring ware.

John Dunhame and Willm Pontus doe undertake to pcure the hering ware repaired and drawne and what they agree for with any that shall doe the worke shalbe payd by the whole Towne according to eich in pporcon of shares.

\* \* \*

At a Townes meeting held xix March 1639.

It is ordered and agreed upon That Thomas Atkins and John Wood shall repaire the hering ware this yeare and shall draw and deliver the herings to eich man according to his shares due to them and shall have ii<sup>s</sup> p thousand [2 shillings per thousand fish] for their paynes of the Towne and after the same rate of the country for those the shalbe allowed to eate and for bayte and to be payd either in money or corne at Harvest at such rate as it doth then passe at from man to man.

It is also agreed upon that whosoever shall take any herings either above or below the ware after the ware is sett or shall robb the ware shall forfaite five for one.

\* \* \*

Att a towne meeting held att the meeting house the 23 of March 1662 It was ordered by the towne that henery Wood and Gorge Bonom with one other whom they shall see meet to be added to them; shall draw or take and devide the herrings to the severall families of the Township of Plymouth whoe shall have theire shares in number according to the number of the psons in theire families and they the saide henery Wood and Gorge Bonum etc are to make meanes for the stopage of the said herrings and takeing of them in theire goeing up att theire owe charge; and they are to lett them goe up on fryday nights[,] on saterday nights and on the Lords daies; and the towne doth prohibite all those that have enterest or shalbee Imployed in the Mill to stopp water when the tide is out of the pond during the time of the herrings; and that they the said pties are hereby authorized to take course for the preventing of Boyes[,] swine and doggs from anoying of them in theire coming up.

*Source*: Town of Plymouth, *Records of the Town of Plymouth*, Vol. I (Plymouth: Avery and Doten, 1889), pp. 3, 52–53.

## DOCUMENT 10: Predator Control and Game Hunting Regulation in Rhode Island Colony (1639, 1646)

Game was generally plentiful when the colonists arrived in New England, and the earliest game laws were directed primarily at controlling livestock predators and limiting the access of the Indians to the wildlife that the colonists wanted for themselves.

### A. Town of Newport, 1639

[A]ll such who shall kill a Fox shall have six shillings and eight pence, for his paines, duly paid unto him by the Treasurer of y^e Towne in which lands it was killed: Provided, that he bring the Head thereof to the said Treasurer; and this order shall be of sufficient authority to the Treasurer to pay and discharge the said summ.

It is further ordered, that all Men who shall kill any Deare (except it be upon his own proper Land) shall bring and deliver half the said Deare into the Tresurie, or pay Forty shillings; and further it is ordered, that the Governour and Deputy Governour shall have authority to give forth a Warrant to some one deputed of each Towne to kill some against the Court times for the Countries use, who shall by his Warrant have Libertie to kill wherever he find; Provided, it be not within any man's enclosure, and to be paid by the Thresurer: Provided, also, that no Indian shall be suffered to kill or destroy at any time or any where.

### B. Town of Portsmouth, 1646

At a meeting, February the 4th, 1646

It is agreed to concur with Newport in an order that there shall be no shootinge of deere for the space of two months; and if any shall shoot, he shall forfeit five pounds; halfe to him that sueth, and the other halfe to the Treasurie. The reason of this order is, that the wolves the more readily come to bayte that they may be catched for the general good of the Island.

. . .

It is ordered, that the wolfe catcher shall be payed out of the treasurie, and that he that killeth a wolfe shall come to Mr. Balston and Mr. Sanford for theire pay.

It is further ordered, that Newport shall pay four pounds for the killinge of a wolfe, and Portsmouth twentie shillings

. . .

It is further ordered, that there shall be noe shootinge of deere from the first of May till the first of November; and if any shall shoot a deere within that time he shall forfeit five pounds; one halfe to him that sueth, and the other to the Treasury.

*Source*: John Russell Bartlett, ed., *Records of the Colony of Rhode Island and Providence Plantations in New England*, Vol. 1: *1636 to 1663* (Providence: A. Crawford Greene, 1856), pp. 84, 85, 113.

---

## DOCUMENT 11: Thomas Hobbes's Social Contract Theory (1651)

In his book *Leviathan*, the English philosopher Thomas Hobbes outlined his social contract theory, positing that people must submit to governmental authority in order to have peace. The treatise, written in support of absolute government and in defense of the acquisition of private property by the monarchy, also offered a coherent statement about the human relationship to the natural world. Jean-Jacques Rousseau, John Locke [see Document 14], and William Blackstone [see Document 18], whose writings were seminal influences on the authors of the U.S. Constitution, took many of their ideas concerning the rights of man, private property, and social contract from *Leviathan*.

The right of nature, which writers commonly call *jus naturale*, is the liberty each man hath, to use his own power, as he will himself, for the preservation of his own nature; that is to say of his own life; and consequently, of doing any thing, which in his own judgment, and reason, he shall conceive to be the aptest means thereunto.

\* \* \*

Whensoever a man transferreth his right or renounceth it; it is either in consideration of some right reciprocally transferred to himself; or for some good he hopeth for thereby. For it is a voluntary act: and of the voluntary acts of every man, the object is some good to himself.

\* \* \*

*[S]uch things as cannot be divided, [should] be enjoyed in common, if it can be; and if the equality of the thing permit, without stint; otherwise proportionally to the number of them that have right.* For otherwise the distribution is unequal, and contrary to equity.

. . . But some things there be, that can neither be divided nor enjoyed in common. Then, the law of nature, which precribeth equity, requireth, *that the entire right; or else, making the use alternate, the first possession to be determined by lot.*

\* \* \*

The final cause, end, or design of men, who naturally love liberty, and dominion over others, in the introduction of that restraint upon themselves, in which we see them live in commonwealths, is the foresight of their own preservation, and of a more contented life thereby; that is to say, of getting themselves out from that miserable condition of war, which is necessarily consequent, as hath been shown . . . , to the natural passions of men, when there is no visible power to keep them in awe, and tie them by fear of punishment to the performance of their covenants, and observation of those laws of nature [discussed earlier].

. . . For the laws of nature, as *justice, equity, modesty, mercy,* and, in sum, *doing to others, as we would be done to,* of themselves, without the terror of some power, to cause them to be observed, are contrary to our natural passions.

\* \* \*

As for the plenty of matter, it is a thing limited by nature, to those commodities, which from the two breasts of our common mother, land and sea, God usually either freely giveth, or for labour selleth to mankind.

For the matter of this nutriment, consisting in animals, vegetals, and minerals, God hath freely laid them before us, in or near to the face of the earth; so as there needeth no more but the labour, and industry of receiving them. Insomuch as plenty dependeth, next to God's favour, merely on the labour and industry of men.

\* \* \*

I find the words *lex civilis,* and *jus civile,* that is to say *law* and *right civile,* promiscuously used for the same thing, even in the most learned authors; which nevertheless ought not to be so. For *right* is *liberty,* namely that liberty which the civil law leaves us: but *civil law* is an *obligation,* which takes from us the liberty which the law of nature gave us. Nature gave a right to every man to secure himself by his own strength, and to invade a suspected neighbour, by way of prevention: but the civil law takes away that liberty.

*Source*: Thomas Hobbes, *Leviathan*, ed. Michael Oakeshott (New York: Macmillan/Collier Books, 1962; original ed. 1651), pp. 103, 105, 121, 129, 185, 214–15.

## DOCUMENT 12: Pollution in Plymouth Colony Harbor (1668)

Pollution from waste disposal, especially in harbors and along wharves, was a problem throughout the colonies. In 1675, for example, Governor Edmund Andros of New York issued a decree forbidding people "to cast dung, dirt or refuse of ye city, or anything to fill up ye harbor or among ye neighbors or neighboring shores, under penalty of forty shillings."[5] But it was not until more than two hundred years later that the U.S. government began to address the issue of harbor pollution [see Document 61]. The records of Plymouth Colony contain one of the first references to harbor pollution in America and probably one of the earliest mentions of any kind of pollution on the continent.

Whereas great complaint is made of great abuse by reason of fishermen that are strangers who fishing on some of the fishing ground on our coast, in Catches dresing and splitting theire fish aboard, through theire Garbdg overboard to the great annoyance of fish which hath any may prove greatly detrementall to the Country; it is ordered by the Court that something be directed from this Court to the Court of the Massachusetts to request them to take some effectuall course for the restraint of such abuse as much as may bee.

*Source*: William Brigham, ed., *The Compact with the Charter and Laws of the Colony of New Plymouth: Together with the Charter of the Council at Plymouth* (Boston: Dutton and Wentworth, 1836), p. 153.

## DOCUMENT 13: William Penn Contracts to Set Aside Timbered Lands (1681)

Whenever an English colony was established, it was customary for the charter holders to set aside a portion of the land for the common use of the whole community. This frequently included wooded land for timber.

In the American colonies, as in all other societies, there were individuals who wanted to use more than their fair share of the common resources [see Document 107] or who wasted common resources through carelessness. Occasionally, when burning timber to clear land for planting, colonists allowed the fires to rage out of control, thereby destroying more woodland than necessary. Others cut an excessive amount of wood. Consequently, just a few years after the establishment of Plymouth Colony, town meetings instituted the first restrictions on the cutting and burning of timber.

As payment for a debt the king had owed to his father, William Penn was granted the territory that is now the state of Pennsylvania. In his contract with the original purchasers and renters of parcels of the land that he received from the English Crown, Penn stipulated that woodland, as well as land to be used for roads, would be set aside prior to the allocation of the individual parcels. Transportation within colonies, between colonies, and to England was deemed essential to the common good, and the provision of wood for shipbuilding was the object of many of the earliest set-asides of timbered land. Sometimes, however, the setting aside of woodland was carried out in order to support unrealistic projects. Penn's scheme to develop mulberry groves as the basis for the establishment of a silk industry proved to be a fanciful notion.

The 11 of July 1681

[1st] That so Soone as it pleaseth God that the abovesaid persons Arrive there, a certaine Quant[ity] of Land or Ground platt shall be laid out for a large Towne or Citty in the most Convenient pla[ce] upon the River for health & Navigation, and every Purchaser & Adventurer shall by lott have soe much Land therein, as will Answer to the Proportion he hath bought or Taken up upon Rent; but it is to be Noted that the Surveyors shall Consider wha[t] Roades or high wayes will be Necessary to the Cittyes, Townes, or through the Lands. Great Roades from Citty to Citty not to Containe Less then Fourty foot in breadth, shall be first laid Out {& declared to be} for highwayes, before the Divident of Acres be laid out for the purchaser, a[nd the] like Observation to be had for Streets in the Townes & Cittyes, that there may be Conven[ient] Roades & Streets preserved not to be Incroached upon by any planter or Builder, and [that] none may build Iregulerly to the dammage of another, in this Custome guide.

* * *

[18ᵗʰ] That in Clearing the Ground, Care be Taken to Leave One Acree of Trees for every five Acres Cleared, especially to Preserve Oak & Mulberries for Silk & Shipping.

Source: *The Papers of William Penn*, Vol. 2: *1680–1684*, ed. Richard S. Dunn and Mary Maples Dunn (Philadelphia: University of Pennsylvania Press, 1982), pp. 98, 100.

---

## DOCUMENT 14: John Locke on Property and Labor (1690)

The English philosopher John Locke, who was greatly influenced by the views of Francis Bacon [see Document 8], Thomas Hobbes [see Document 11], and René Descartes, was in turn a source for the drafters of the U.S. Constitution and the early members of Congress. His thesis that an individual's property should not exceed the amount of property that a laborer could tend was later incorporated into the Homestead Act [see Document 42].

[I]f gathering the Acorns, or other Fruits of the Earth, &c. makes a Right to them, then any one may *ingross* as much as he will. To which I Answer, Not so. The same Law of Nature, that does by this means give us Property, does also *bound* that *Property* too. *God has given us all things richly*, 1 Tim. vi. 12. is the Voice of Reason confirmed by Inspiration. But how far has he given it us? *To enjoy.* As much as any one can make use of to any Advantage of Life before it spoils; so much he may by his labour fix a Property in. Whatever is beyond this, is more than his Share, and belongs to others. Nothing was made by God for Man to spoil or destroy. And thus considering the Plenty of natural Provisions there was a long time in the World, and the few Spenders; and to how small a Part of that Provision the Industry of one Man could extend it self, and ingross it to the Prejudice of others; especially keeping within the *Bounds* set by Reason *of* what might serve for his *Use*, there could be then little room for Quarrels or Contentions about Property so establish'd.

But the *chief Matter of Property* being now not the Fruits of the Earth, and the Beasts that subsist on it, but *the Earth it self*; as that which takes in and carries with it all the rest: I think it is plain, that *Property* in that too is acquir'd as the former. *As much Land* as a Man Tills, Plants, Improves, Cultivates, and can use the Product of, so much is his *Property*. He by his Labour does, as it were, inclose it from the Common. . . .

Nor was this *Appropriation* of any parcel of *Land*, by improving it, any Prejudice to any other Man, since there was still enough, and as good left; and more than the yet unprovided could use. . . .

God gave the World to Men in Common; but since he gave it them for their Benefit, and the greatest conveniencies of Life they were capable to draw from it, it cannot be supposed he meant it should always remain common and uncultivated. He gave it to the use of the industrious and rational. . . .

'Tis true, in *Land* that is *common* in *England*, or any other Country, where there is plenty of People under Government, who have Money and Commerce, no one can inclose or appropriate any part, without the consent of all his Fellow-Commoners: Because this is left common by Compact, *i.e.* by the Law of the Land, which is not to be violated. And tho' it be common, in respect of some Men, it is not so to all Mankind; but is the joint property of this Country, or this Parish. Besides, the remainder, after such Inclosure, would not be as good to the rest of the Commoners as the whole was, when they could all make use of the whole; whereas in the Beginning and first peopling of the great Common of the World, it was quite otherwise.

\* \* \*

Nor is it so strange, as perhaps before consideration it may appear, that the *Property of Labour* should be able to over-balance the Community of Land. For 'tis *Labour* indeed that *puts the difference of value* on every thing.

\* \* \*

Land that is left wholly to Nature, that hath no improvement of Pasturage, Tillage, or Planting, is called, as indeed it is, *waste*; and we shall find the benefit of it amount to little more than nothing.

This shews, how much numbers of Men are to be preferred to largenesse of Dominions; and that the increase of Lands and the right imploying of them is the great Art of Government.

*Source*: John Locke, "Of Civil Government" in *Two Treatises of Government*, in *The Works of John Locke esq.*, Vol. 2 (London: John Churchill, 1714), pp. 167–68, 170.

---

## DOCUMENT 15: John Ray on Gardens and Wilderness (1691)

The Father of English Natural History, John Ray associated a cultivated landscape with civilization, and wilderness with barbarism. He believed that the richness and diversity of nature was a reflection of God's magnificence and that God had provided this great bounty for human use. Most of Ray's writings were devoted to the systematic description of plants and animals.

I perswade my self, that the bountiful and gracious Author of Man's Being and Faculties, and all Things else, delights in the Beauty of his Creation, and is well pleased with the Industry of Man, in adorning the Earth with beautiful Cities and Castles; with pleasant Villages and Country-Houses; with regular Gardens and Orchards, and Plantations of all Sorts of Shurbs, and Herbs, and Fruits, for Meat, Medicine, or moderate Delight; with shady Woods and Groves, and Walks set with Rows of elegant Trees; *with Pastures cloathed with Flocks, and Valleys covered over with Corn,* and Meadows burthened with Grass, and whatever else differenceth a civil and well cultivated Region, from a barren and dissolate Wilderness.

If a Country thus planted and adorn'd, thus polished and civilized, thus improved to the height by all manner of Culture for the Support and Sustenance, and convenient Entertainment of innumerable multitudes of People, be not to be preferred before a barbarous and inhospitable *Scythia,* without Houses, without Plantations, without Corn-fields or Vineyards, where the roving *Hords* of the savage and truculent Inhabitants, transfer themselves from place to place in Waggons, as they can find Pasture and Forrage for their Cattle, and live upon Milk, and Flesh roasted in the Sun, at the Pomels of their Saddles; or a rude and unpolished *America,* peopled with slothful and naked *Indians,* instead of well-built Houses, living in pitiful Huts and Cabbins, made of Poles set end-ways; then surely the brute Beasts Condition, and manner of Living, to which, what we have mention'd doth nearly approach, is to be esteem'd better than Man's, and Wit and Reason was in vain bestowed on him.

*Source*: John Ray, *The Wisdom of God Manifested in the Works of the Creation*, 8th ed. (London: William and John Innys, 1722), pp. 164–65.

---

## DOCUMENT 16: Jonathan Edwards on God and Nature (1739)

Like John Ray, the New England theologian Jonathan Edwards attempted to show how nature—the totality of the many aspects of the creation—was a reflection of God's magnificence. Despite his conviction that the whole natural world was created for human use, Edwards believed that people had to live in harmony with the natural world if they were to enjoy its benefits and bask in God's love. According to Edwards, the contemplation of nature helped to turn people away from evil action. Unlike Hobbes [see Document 11], Edwards concluded that by using reason, people could deduce civil and moral precepts from the "nature of things." Edwards' writings and sermons set the groundwork for a Protestant aesthetic spirituality (a belief that the experience of beauty and nature made one sensitive to the presence of God) that was popularized by the transcendentalists in the nineteenth century [see Documents 36, 44] and found its way into the early writings of John Muir [see Document 47].

The sense I had of divine things, would often of a sudden kindle up, as it were, a sweet burning in my heart; an ardor of soul, that I know not how to express.

Not long after I first began to experience these things . . . I walked abroad alone, in a solitary place in my father's pasture, for contemplation. And as I was walking there, and looking up on the sky and clouds, there came into my mind so sweet a sense of the glorious *majesty* and *grace* of God, that I know not how to express. I seemed to see them both in a sweet conjunction; majesty and meekness joined together; it was a sweet, and gentle, and holy majesty; and also a majestic meekness; an awful sweetness; a high, and great, and holy gentleness.

After this my sense of divine things gradually increased, and became more and more lively, and had more of that inward sweetness. The appearance of every thing was altered; there seemed to be, as it were, a calm, sweet cast, or appearance of divine glory, in almost every thing. God's excellency, his wisdom, his purity and love, seemed to appear in every thing; in the sun, moon, and stars; in the clouds and blue sky; in the grass, flowers, trees; in the water, and all nature.

*Source*: Jonathan Edwards, "Personal Narrative," in Clarence Faust and Thomas Johnson, eds., *Jonathan Edwards: Representative Selections* (New York: American Book Company, 1935), pp. 60–61.

## DOCUMENT 17: Peter Kalm on Land Management (1753)

Peter Kalm was a professor of economy at the University of Aobo in Swedish Finland. His travels in America between 1748 and 1751 were one of a series of voyages sponsored by the Royal Academy at Stockholm "to make . . . such observations and collections of seeds and plants as would improve the *Swedish* husbandry, gardening, manufactures, arts and sciences. [The great Swedish botanist] Dr. [Carl] Linnaus . . . thought that a voyage through *North America* would be yet of a more extensive utility, than that through [Siberia and Iceland]; for the plants of America were then little known, and not scientifically de-scribed."[6] Here Kalm discusses land use in the colonies.

The rye grows very poorly in most of the fields, which is chiefly owing to the carelessness in agriculture, and to the poorness of the fields, which are seldom or never manured. After the inhabitants have converted a tract of land into a tillable field, which had been a forest for many cen-turies together, and which consequently had a very fine soil, they use it as such as long as it will bear any corn; and when it ceases to bear any, they turn it into pasture for the cattle, and take new corn-fields in another place, where a fine soil can be met with and where it has never been made use of for this purpose. This kind of agriculture will do for some time; but it will afterwards have bad consequences, as every one may clearly see. A few of the inhabitants, however, treat their fields a little better: the *English* in general have carried agriculture to a higher degree of perfection than any other nation. But the depth and richness of the soil, which those found here who came over from England (as they were preparing land for plowing, which had been covered with woods from times immemorial) misled them, and made them careless husbandmen. It is well known, that the *Indians* lived in this country for several cen-turies before the *Europeans* came into it; but it is likewise known, that they lived chiefly by hunting and fishing, and had hardly any fields. They planted maize, and some species of beans and gourds; and at the same time it is certain, that a plantation of such vegetables as serve an *Indian* family during one year, take up no more ground than a farmer in our country takes to plant cabbage for his family upon; at least, a farmer's cabbage and turnep ground, taken together, is always as exten-sive, if not more so, than the corn-fields and kitchen-gardens of an *Indian* family. Therefore, the *Indians* could hardly subsist for one month upon the produce of their gardens and fields. Commonly, the little villages of *Indians* are about twelve or eighteen miles distant from each other. From hence one may judge, how little ground was formerly employed for planting corn-fields; and the rest was overgrown with thick and tall trees. And though they cleared (as is yet usual) new ground, as soon as the old one had quite lost its fertility; yet such little pieces as they made use of were very inconsiderable, when compared to the vast forest which remained. Thus the upper fertile soil increased considerably for centuries together; and the *Europeans* coming to *America* found a rich, fine soil before them, lying as loose between the trees as the best bed in a garden. They had nothing to do but to cut down the wood, put it up in heaps, and to clear the dead leaves away. They could then immediately proceed to plowing, which in such loose ground is very easy; and having sown their corn, they got a most plentiful harvest. This easy method of getting a rich crop has spoiled the *English* and other *European* settlers, and induced them to adopt the same method of agriculture which the *Indians* make use of; that is, to sow uncultivated grounds, as long as they will produce a crop without manuring, but to turn them into pastures as soon as they can bear no more, and to take on new spots of ground, covered since time immemorial with woods, which have been spared by the fire or the hatchet ever since the creation.

*Source*: Peter Kalm, *Travels into North America; Containing Its Natural History, and a Circumstantial Account of Its Plantations and Agriculture in General*, trans. John Reinhold Forster, Vol. 2 (Warrington, England: William Eyres, 1772), pp. 191–94.

## DOCUMENT 18: William Blackstone's *On the Rights of Things* (1765–1769)

William Blackstone's history of the doctrines of English law greatly influenced the development of jurisprudence in the United States, and the roots of American law concerning private property rights can be traced to this seminal work.

There is nothing which so generally strikes the imagination, and engages the affections of mankind, as the right of property; or that sole and despotic dominion which one man claims and exercises over the external things of the world, in total exclusion of the right of any other individual in the universe. And yet there are very few, that will give themselves the trouble to consider the original and foundation of this right. . . .

In the beginning of the world, we are informed by holy writ, the all-bountiful creator gave to man, "dominion over all the earth; and over the fish of the sea, and over the fowl of the air, and over every living thing that moveth upon the earth." This is the only true and solid foundation of man's dominion over external things, whatever airy metaphysical notions may have been started by fanciful writers upon this subject. The earth therefore, and all things therein, are the general property of all mankind, exclusive of other beings, from the immediate gift of the creator. And, while the earth continued bare of inhabitants, it is reasonable to suppose, that all was in common among them, and that every one took from the public stock to his own use such things as his immediate necessities required.

These general notions of property were then sufficient to answer all the purposes of human life; and might perhaps still have answered them, had it been possible for mankind to have remained in a state of primaeval simplicity: as may be collected from the manners of many American nations when first discovered by the Europeans; and from the antient method of living among the first Europeans themselves. . . . Not that this communion of goods seems ever to have been applicable, even in the earliest ages, to ought but the *substance* of the thing; nor could it be extended to the *use* of it. For, by the law of nature and reason, he who first began to use it, acquired therein a kind of transient property, that lasted so long as he was using it, and no longer. . . .

But when mankind increased in number, craft, and ambition, it became necessary to entertain conceptions of more permanent dominion; and to appropriate to individuals not the immediate *use* only, but the very *substance* of the thing to be used. Otherwise innumerable tumults must have arisen, and the good order of the world been continually broken and disturbed. . . .

* * *

Property, both in lands and moveables, being thus originally acquired by the first taker, which taking amounts to a declaration that he intends to appropriate the thing to his own use, it remains in him, by the principles of universal law, till such time as he does some other act which shews an intention to abandon it: for then it becomes, naturally speaking, *publici juris* once more, and is liable to be again appropriated by the next occupant. . . .

But this method of one man's abandoning his property, and another's seising the vacant possession, however well founded in theory, could not long subsist in fact. It was calculated merely for the rudiments of civil society, and necessarily ceased among the complicated interests and artificial refinements of polite and established governments. In these it was found, that what became inconvenient or useless to one man was highly convenient and useful to another; who was ready to give in exchange for it some equivalent, that was equally desirable to the former proprietor.

*Source*: William Blackstone, *Commentaries on the Laws of England*, Vol. 2 (Philadelphia: Bell, 1771–1772), pp. 2–4, 9.

---

## DOCUMENT 19: John Bartram on Reclaiming Florida's Wetlands (1767)

---

Traditionally, the marsh was a wild, evil place that was feared and had to be subdued. It was thought to be the home of murderous spirits and dangerous monsters such as Grendel, the huge moor-stalking subject of the tenth century Anglo-Saxon epic poem *Beowulf* "who held the wasteland, fens, and marshes."[7]

The colonists brought this view of wetlands with them to the New World. In the United States, until the latter part of the twentieth century, marshes and other wetlands were generally regarded as wastelands, evil places, and sources of diseases such as malaria; they were places to be avoided and, if technologically possible, to be done away with.

John Bartram was the first American-born naturalist. Although he was self-educated and never traveled to Europe, Bartram knew all the major European naturalists of his day either personally—because they came to visit his acclaimed garden at his home outside Philadelphia (for example, Peter Kalm [see Document 17])—or through written correspondence. Eager to have contact with the greatest thinkers of his time, Bartram urged his friend and fellow Philadelphian Ben Franklin to organize a society of the "most ingenious and curious men"[8] in America, and in 1692 Franklin formed the American Philosophical Society, the first scientific organization in America, to promote useful knowledge and encourage scientific agriculture.

When Bartram was in his sixties, he was appointed botanist to the king of England, and in this capacity he surveyed the newly acquired territory of Florida for the English Crown. Although he remarked on the richness of the wetlands and recognized their function as a sanctuary for young fish, he nevertheless was the first of many people to recommend the reclamation of Florida's wetlands.

The pine-lands, as they are here called, contain a variety of soil, according to their different situations. . . . The pine-land, by the help of dung and cultivation, will produce good corn, potatoes, and cotton; the large palmetto declining ground, between the pines and swamps, are moist and seem rich, and perhaps will suit both corn and indigo; but the shelly bluffs seem to be the most fertile spots of high ground, and the Indians chief plantations for corn and pumkins: That which is called hammocky ground is generally full of large evergreens and water-oaks, mixed with red-bay and magnolia, and in many places the great palmetto or cabbagetree: this is generally reckoned proper both for corn, cotton, and indigo: but the marshes and swamps (so extensive upon the river St. John's) are exceedingly rich, the last of which are full of large ash, maple, and elm, being of an unknown depth of rich mud; so are the marshes on the upper part of the river, which are covered with water-canes and reeds, as the lower marshes are with grass and weeds; all of which when they are drained dry, will produce, in all probability, great crops of corn and indigo, and without much or any draining, a fine increase of rice. . . .

St. John's river, by its near affinity to the sea, is well replenished with variety of excellent fish, as bass, sea-trout, sheep-head, drums, mullets, cats, garr, sturgeon, stingrays; and near its mouth, oysters, crabs, and shrimps, sharks and porpoises, which will doubtless continue. . . . Its shores, being generally shoal . . . afford a fine asylum to the young fry against their devouring enemies.

*Source*: William Stork, *A Description of East-Florida, with a Journal, Kept by John Bartram of Philadelphia, Botanist to His Majesty for the Floridas; upon a Journey from St. Augustines up the River St. John, as Far as the Lakes*, 4th ed. (London: Faden & Jefferys, 1774), p. 34.

# Part II

# Politicians, Naturalists, and Artists in the New Nation, 1776–1840

On the eve of the Revolution, the thirteen belligerent British colonies clustered along the Atlantic coast had a population of approximately 2.5 million, including about half a million blacks, and approximately 90 percent of the workforce was engaged in agriculture. By 1790 there were nearly 4 million people in the United States, by 1820 over 9.5 million, and by 1830 close to 13 million. In just sixty-five years the nation's population had more than quintupled. It had also begun to be both more urban and more industrial.

Educated Americans, such as Thomas Jefferson and James Madison, who followed trends in Europe and purchased many of their reading materials abroad, were knowledgeable about and interested in the theories of European economists, including Thomas Malthus, concerning the relationship of populousness, landownership, and poverty [see Documents 21 and 26]. However, these issues did not arouse the passions of most Americans. Although there were poor people in the United States, large pockets of urban poverty comparable to those found in London and Paris were unknown in the United States until well into the nineteenth century. Furthermore, in late eighteenth- and early nineteenth-century America, there was plenty of unsettled land available for farmers—unlike in western Europe, where much of the arable land had been farmed for generations. Indeed, the U.S. government was anxious to populate the nation's western lands with Europeans and their descendants.

## GROWTH OF THE NATION

In the Treaty of 1783, Great Britain not only recognized the independence of the thirteen rebellious colonies but also, very generously, established the Mississippi River as the western boundary of the new nation, far beyond the westernmost border of any of the thirteen colonies. Twenty years later, Thomas Jefferson's purchase of the Louisiana Territory from France moved the young nation's western boundary to the Rocky Mountains and incorporated into the country the Great Plains, a terrain vastly different from the deciduous forest regions of the Northeast.

The territories west of the original thirteen colonies gained few inhabitants during the first half-century of the country's existence. One of the most formidable obstacles to the western movement of the populace was the difficulty of overland transportation. Even within the former colonies, overland travel was slow and arduous because there were virtually no paved roads. Indeed, the nation's earliest cities—Boston, New York, Philadelphia, and Charleston—were on the Atlantic coast, and the first inland areas to develop had easy river access to the Atlantic, since boat transportation was far more efficient than road travel.

The Indians—abetted at times by Britain, France, and Spain—also posed a continuous hindrance to the expansion of the country. In 1829 President Andrew Jackson proposed that the remaining tribes in the eastern United States be bodily moved and resettled west of the Mississippi; this was the beginning of an official U.S. policy of removing Indians from lands that whites wanted. In 1832, Indians, under the leadership of Black Hawk, tried to reoccupy their land in Illinois, but their rebellion was crushed [see Document 32] in a bloody battle. In Florida, which the United States had acquired from Spain in 1819, the Seminole Indians also fiercely resisted eviction and white settlement, but by the end of the 1830s three-quarters of them had been moved to Oklahoma.

The extensive territorial growth of the country, along with its increasing industrialization, demanded the development of an efficient transportation infrastructure, and by the late 1830s several hard-surfaced turnpikes had been constructed, steamboats were plying some of the nation's rivers, railroads had begun to radiate out from major cities, and the Erie Canal had been completed, connecting the Hudson River and the East with Lake Erie and the West. Simultaneously, the occupation base of the country was changing; by the 1830s less than three-quarters of the nation's workforce was employed on farms. It was a trend that had disturbed Americans like Thomas Jefferson when it first became apparent, but as the century progressed, the increasing industrialization and urbanization of the United States was recognized as necessary for national economic development [see Document 20].

Since the end of the seventeenth century, trappers and traders eager to obtain furs to ship to Europe had been slowly moving into the

the navy [see Document 28]. In 1828 President John Quincy Adams took over a timbered area of Pensacola, Florida, to establish a naval station and then set aside 30,000 acres of live-oak forest land, on Santa Rosa Island, near Pensacola, to provide wood for the future shipbuilding needs of the navy.[1]

Not much thought was given to conservation at the state level either. But in 1818 the nation's first bird protection law [see Document 29] was passed in Massachusetts because farmers recognized that birds provided not only meat and feathers but also some measure of crop protection.

During these early years of the country, naturalists such as William Bartram [see Document 23] and John James Audubon [see Document 33] and explorers such as Meriwether Lewis and William Clark [see Document 27] traveled through the United States, drawing, painting, and inventorying the nation's flora, fauna, and geological resources. Their reports and journals extolled the beauty and diversity of the land and the creatures that inhabited it. But they also depicted the effects of the human presence on the natural landscape; by the 1830s writers were deploring the destruction of the buffalo, the killing of birds, and the wanton slaughter of many species of wild animals.

Other writers and artists increasingly conveyed a growing appreciation of the wilderness, often accompanied by a fascination with American Indians and their way of life. Probably the earliest expression of this admiration for both wild nature and Native Americans is to be found in the myth of the noble savage who loves and understands the wilderness and is unhappy when forced to live apart from it. The myth, whose origins date back to the seventeenth century, was in part a reaction to growing industrialization—a look backward to a disappearing way of life—and in part a rejection of the pastoral ideal. It was the springboard for countless poems, stories, and paintings, ranging from Philip Freneau's "Indian Student" [see Document 22] to George Catlin's realistic portraits of Indian chiefs. Catlin, who traveled to the West to paint Indians in their natural settings, fell in love with the magnificent western scenery and, realizing that the land would soon be overrun by settlers from the East, suggested that some of the grasslands be set aside for a national park [see Document 31]. The novelist James Fenimore Cooper also saw the hand of progress destroying the old way of life. The hero of his Leatherstocking series is an upstate New York frontiersman with a great attachment to the wilderness and a disdain for many of the laws and trappings of modern civilization [see Document 30].

interior of the country and encouraging Indians, often ag
to join them in hunting and trapping in the furtherance (
than for mere subsistence. Miners and, later, loggers anc
lowed the trappers and traders west. In 1720 the first lead
country was opened in Missouri. As roads and trains imp
trickle of pioneers going west became a flood.

## CARING FOR THE LAND AND ITS RESOURCES

The abundance of cheap land worked against the developn
an attitude of respect for the land and its resources. Early Am
farmers found it more expedient to deplete the land and move on
to maintain their property. Even farmers such as George Washin
and Thomas Jefferson, who gave a great deal of thought to the car
their land, recognized that the availability of cheap land made it i
practical to spend large sums for fertilizer for their fields. It was on
when land values had risen sufficiently to warrant large expenditure
that farmers began to devote energy and money to the maintenance o
the soil's fertility [see Document 25].

Forests were cleared for farmland with great abandon. Except in the
few instances where efforts were made to maintain stands of trees for
the shipbuilding industry or to satisfy local construction and firewood
needs [see Document 13], little serious thought was given to the importance of woodland preservation. Even Dr. Benjamin Rush's call in
1791 for the preservation of the sugar maple [see Document 24] may
have been motivated more by social and political considerations than
by environmental factors.

At the federal level there was little evidence of concern about resources and their conservation during the first half-century of the nation's existence. The acquisition of new lands, access to Atlantic
fisheries, and the availability of timber for shipbuilding were the primary resource issues. The Treaty of 1783, the Treaty of Ghent (which
settled the War of 1812 and which John Quincy Adams helped to
negotiate), and the Treaty of 1818 all stipulated that Americans would
continue to have access to the fisheries off the coast of Newfoundland.

In 1799, Congress, conscious of the need for suitable timber for tall
ship masts and also aware that hardwoods were being stolen from public lands along the coast of the Gulf of Mexico, appropriated $200,000
for President John Adams to purchase two heavily forested islands off
the Georgia coast, Grover's Island and Blackbeard's Island. The policy
of protecting timber resources for the fledgling U.S. Navy was continued under the administrations of James Monroe and John Quincy Adams, during which time Congress passed three acts—in 1817, 1822,
and 1827—to set aside land along the Gulf Coast to provide timber for

## DOCUMENT 20: Thomas Jefferson on Agrarianism and Industrialization (1785, 1816)

Thomas Jefferson's early agrarianism reflected a conviction that contact with the land improves the human soul and is good for the human spirit, but eventually Jefferson recognized the importance of industrial development for the well-being of the United States. Later in the century, however, as the Industrial Revolution began to affect the lives of most Americans, the transcendentalists, including Ralph Waldo Emerson and Henry David Thoreau [see Documents 36 and 44], would once again raise the banner of agrarianism.

### A. From *Notes on the State of Virginia*, 1785

The political oeconomists of Europe have established it as a principle that every state should endeavour to manufacture for itself; and this principle, like many others, we transfer to America, without calculating the difference of circumstance which should often produce a difference of result. In Europe the lands are either cultivated, or locked up against the cultivator. Manufacture must therefore be resorted to of necessity not of choice, to support the surplus of their people. But we have an immensity of land courting the industry of the husbandman. Is it best then that all our citizens should be employed in its improvement, or that one half should be called off from that to exercise manufactures and handicraft arts for the other? Those who labour in the earth are the chosen people of God, if ever He had a chosen people, whose breasts He has made His peculiar deposit for substantial and genuine virtue. It is the focus in which he keeps alive that sacred fire, which otherwise might escape from the face of the earth. Corruption of morals in the mass of cultivators is a phenomenon of which no age nor nation has furnished an example. . . . While we have land to labour then, let us never wish to see our citizens occupied at a work-bench, or twirling a distaff. Carpenters, masons, smiths, are wanting in husbandry; but, for the general operations of manufacture, let our work-shops remain in Europe.

### B. To Benjamin Austin, January 9, 1816

We must now place the manufacturer by the side of the agriculturalist. . . . The grand inquiry now is, shall we make our own comforts, or go without them at the will of a foreign nation? He, therefore, who is now against domestic manufacture, must be for reducing us either to dependence on that foreign nation, or to be clothed in skins, and to live like wild beasts in dens and caverns. I am not one of these. Experience has taught me that manufactures are now as necessary to our independence as to our comfort.

*Source*: **A**. Thomas Jefferson, *Notes on the State of Virginia* (London: Stockdale, 1787), pp. 273–75. **B**. Thomas Jefferson Randolph, ed., *Memoirs, Correspondence, and Private Papers of Thomas Jefferson*, Vol. 4 (London: Colburn and Bentley, 1829), p. 279.

---

## DOCUMENT 21: James Madison on Population and Property (1786, 1788)

---

One of the drafters of the U.S. Constitution, James Madison devoted much thought to the issue of private property ownership. Among his concerns were how an increase in the population of the country would affect the equitable distribution of land and the divisions that would develop as a result of ownership of property or the lack thereof. Madison and many others in both Europe and the United States were interested in these issues well before Robert Malthus published his famous *Essay on the Principles of Population* [see Document 26].

### A. To Thomas Jefferson, June 19, 1786

I have no doubt but that the misery of the lower classes will be found to abate wherever the Government assumes a freer aspect, & the laws favor a subdivision of property, yet I suspect that the difference will not fully account for the comparative comfort of the mass of people in the United States. Our limited population has probably as large a share in producing this effect as the political advantages which distinguish us. A certain degree of misery seems inseparable from a high degree of populousness. If the lands in Europe which are now dedicated to the amusement of the idle rich, were parcelled out among the idle poor, I readily conceive the happy revolution which would be experienced by a certain proportion of the latter. But still would there not remain a great proportion unrelieved? No problem in political economy has appeared to me more puzzling than that which relates to the most proper distribution of the inhabitants of a country fully peopled. Let the lands be shared among them ever so wisely, & let them be supplied with labourers ever so plentifully; as there must be a great surplus of subsistence, there will also remain a great surplus of inhabitants, a greater by far than will be employed in cloathing both themselves & those who feed them, and in administering to both, every other necessary & even comfort of life. What is to be done with this surplus? Hitherto we have seen them distributed into manufactures of superfluities, idle proprietors of productive lands, domestics, soldiers, merchants, mariners, and a few other less numerous classes. All these classes notwithstanding have been found insufficient to absorb the redundant members of a populous society; and yet a reduction of most of those classes enters into the very reform which appears so necessary & desirable. From a more equal partition of property, must result a greater simplicity of manners, consequently a less consumption of manufactured superfluities, and a less proportion of idle proprietors & domestics.

### B. From *The Federalist*, Number 10, 1788

As long as the reason of man continues fallible, and he is at liberty to exercise it, different opinions will be formed. As long as the connexion subsists between his reason and his self-love, his opinions and his passions will have a reciprocal influence on each other; and the former will be objects to which the latter will attach themselves. The diversity in the faculties of men, from which the rights of property originate, is not less an insuperable obstacle to a uniformity of interests. The protection of these faculties is the first object of government. From the protection of different and unequal faculties of acquiring property, the possession of different degrees and kinds of property immediately results; and from the influence of these on the sentiments and view of the respective proprietors, ensues a division of the society into different interests and parties.

. . . [T]he most common and durable source of factions has been the various and unequal distribution of property—Those who had and those who are without property have ever formed distinct interests in society. Those who are creditors, and those who are debtors, fall under a like discrimination. A landed interest, a manufacturing interest, a mercantile interest, a moneyed interest, with many lesser interests, grow up of necessity in civilized nation, and divide them into different classes, actuated by different sentiments and views. The regulation of these various and interfering interests forms the principal task of modern legislation, and involves the spirit of party and faction in the necessary and ordinary operations of the government.

*Source*: **A**. James Madison, letter to Thomas Jefferson, June 19, 1786, in Saul K. Padover, ed., *The Complete Madison: His Basic Writings* (New York: Harper, 1953), pp. 317–18. **B**. Alexander Hamilton, James Madison and John Jay, *The Federalist, on the New Constitution* (New York: Williams and Whiting, 1810), pp. 62–64.

## DOCUMENT 22: Philip Freneau's Noble Savage (1788)

The Swiss-born American poet and essayist Philip Freneau used the romantic myth of the noble savage as the basis for his tale of the Indian student. By juxtaposing the life of a Harvard student with a life lived closer to nature, Freneau was subtly raising a question about the superiority of Western values, a question that would be posed more vociferously by environmentalists in the latter part of the twentieth century. The student, who yearns for a return to wild nature and lays aside his Virgil, embodies the romantic rejection of the pastoral ideal of the cultivated landscape [see Document 2].

From Susquehanna's utmost springs
Where savage tribes pursue their game,
His blanket tied with yellow strings,
A shepherd of the forest came.

Not long before, a wandering priest
Express'd his wish, with visage sad—
"Ah, why (he cry'd) in Satan's Waste,
"Ah, why detain so fine a lad?

"In Yanky land there stands a town
"Where learning may be purchas'd low—
"Exchange his blanket for a gown,
"And let the lad to college go."—

From long debate the Council rose,
And viewing *Shalum's* tricks with joy,
To *Harvard hall*, o'er wastes of snows,
They sent the copper-colour'd boy.

                        * * *

Awhile he writ, awhile he read,
Awhile he learn'd the grammar rules—
An Indian savage so well bred
Great credit promis'd to their schools.

Some thought he would in *law* excel,
Some said in *physic* he would shine;
And one that knew him, passing well,
Beheld, in him, a sound divine.

But those of more discerning eye
Even then could other prospects show,

And saw him lay his *Virgil* by
To wander with his dearer *bow*.

The tedious hours of study spent,
The heavy-moulded lecture done,
He to the woods a hunting went,
But sigh'd to see the setting sun.

No mystic wonders fir'd his mind;
He sought to gain no learn'd degree,
But only sense enough to find
The squirrel in the hollow tree.

* * *

"And why (he cry'd) did I forsake
"My native wood for gloomy walls;
"The silver stream, the limpid lake
"For musty books and college halls.

"A little could my wants supply—
"Can wealth and honour give me more;
"Or, will the sylvan god deny
"The humble treat he gave before?

* * *

"Where Nature's ancient forests grow,
"And mingled laurel never fades,
"My heart is fix'd;—and I must go
"To die among my native shades."

He spoke, and to the western springs,
(His gown discharg'd, his money spent)
His blanket tied with yellow strings,
The shepherd of the forest went.

*Source*: Philip Freneau, "The Indian Student," *The Poems and Miscellaneous Works*, ed. Lewis Leary (Delmar, NY: Scholar, 1975; facsimile of two works printed separately in 1786 and 1788 by F. Bailey), pp. 69–71.

## DOCUMENT 23: William Bartram on the Human Impact on the Environment (1791)

Before setting out on his own to explore the southeastern coast of America, the renowned botanical illustrator William Bartram had traveled through much of the area with his father, John Bartram [see Document

19] and had attempted to raise indigo (a plant used to obtain a blue dye) in Florida. In his travels along the St. Johns River, William Bartram took note of the interplay between people—both Indians and whites—and their surroundings, remarking on how people changed the face of the land to suit their own particular needs when they settled in a particular place, and on how humans could turn a beautiful place into a devastated landscape. Many people consider William Bartram to be the first American environmentalist.

About the middle of May, every thing being in readiness to proceed up the river, we sat sail. . . .

We had a pleasant day, the wind fair and moderate, and ran by Mount Hope, so named by my father John Bartram, when he ascended this river, about fifteen years ago. It is a very high shelly bluff, upon the little lake. It was at that time a fine Orange grove, but now cleared and converted into a large indigo plantation, the property of an English gentleman, under the care of an agent. In the evening we arrived at Mount Royal, where we came to and stayed the night. . . .

From this place we enjoyed a most enchanting prospect of the great Lake George, through a grand avenue, if I may so term this narrow reach of the river, which widens gradually for about two miles, towards its entrance into the lake, so as to elude the exact rules of perspective, and appears of an equal width.

At about fifty yards distance from the landing place, stands a magnificent Indian mount. About fifteen years ago I visited this place, at which time there were no settlements of white people, but all appeared wild and savage; yet in that uncultivated state it possessed an almost inexpressible air of grandeur, which was now entirely changed. At that time there was a very considerable extent of old fields round about the mount; there was also a large orange grove, together with palms and live oaks, extending from near the mount, along the banks, downwards, all of which has since been cleared away to make room for planting ground. But what greatly contributed towards completing the magnificence of the scene, was a noble Indian highway, which led from the great mount, on a straight line, three quarters of a mile, first through a point or wing of the orange grove, and continuing thence through an awful forest of live oaks, it was terminated by palms and laurel magnolias, on the verge of an oblong artificial lake, which was on the edge of an extensive green level savanna. This grand highway was about fifty yards wide, sunk a little below the common level, and the earth thrown up on each side, making a bank of about two feet high. Neither nature nor art could any where present a more striking contrast, as you approached this savanna. The glittering water pond played on the sight, through the dark grove, like a brilliant diamond, on the bosom of the illumined savanna, bordered with various flowery shrubs and plants; and as we advanced into the plain, the sight was agreeably relieved by a distant view of the forests, which partly environed the green expanse on the left hand, whilst the imagination was still flattered and entertained by the far distant misty points of the surrounding forests, which projected into the plain, alternately appearing and disappearing, making a grand sweep round on the right, to the distant banks of the great lake. But that venerable grove is now no more. All has been cleared away and planted with indigo, corn, and cotton, but since deserted: there was now scarcely five acres of ground under fence. It appeared like a desart to a great extent, and terminated, on the land side, by frightful thickets, and open pine forests.

*Source*: William Bartram, *Travels Through North and South Carolina, Georgia and Florida* (Philadelphia: James and Johnson, 1791), pp. 96–98.

## DOCUMENT 24: Benjamin Rush on Saving the Sugar Maple (1791)

The relationship between conservation and political and social thought is evident in this letter by a prominent Revolutionary era physician. It combines a plea for the preservation of the sugar maple with a quest for the emancipation of slaves.

The *Acer Sacharinum* of Linnaeus, or Sugar Maple-tree, grows in great quantities in the western counties of all the Middle States of the American Union. Those which grow in New-York and Pennsylvania yield the sugar in a greater quantity than those which grow on the waters of the Ohio.

\* \* \*

But the profit of the maple tree is not confined to its sugar. It affords a most agreeable molasses, and an excellent vinegar. The sap which is suitable for these purposes is obtained after the sap which affords the sugar has ceased to flow, so that the manufactories of these different products of the maple tree, by *succeeding*, do not interfere with each other. The molasses may be made to compose the basis of a pleasant summer beer. The sap of the maple is moreover capable of affording a spirit, but we hope this precious juice will never be prostituted by our citizens to this ignoble purpose. Should the use of sugar in diet become more general in our country, it may tend to lessen the inclination or supposed necessity for spirits, for I have observed a relish for sugar in diet to be seldom accompanied by a love for strong drink.

* * *

Cases may occur in which sugar may be required in medicine, or in diet, by persons who refuse to be benefited, even indirectly by the labour of slaves. In such cases, the innocent maple sugar will always be preferred.

It has been said, that sugar injures the teeth, but this opinion now has so few advocates, that it does not deserve a serious refutation.

To transmit to future generations, all the advantages which have been enumerated from the maple tree, it will be necessary to protect it by law, or by a bounty upon the maple sugar, from being destroyed by the settlers in the maple country, or to transplant it from the woods, and cultivate it in the old and improved parts of the United States. An orchard consisting of 200 trees, planted upon a common farm would yield more than the same number of apple trees, at a distance from a market town. A full grown tree in the woods yields five pounds of sugar a year. If a greater exposure of a tree to the action of the sun, has the same effects upon the maple, that it has upon other trees, a larger quantity of sugar might reasonably be expected from each tree planted in an orchard.

* * *

In contemplating the present opening prospects in human affairs, I am led to expect that a material share of the happiness, which Heaven seems to have prepared for a part of mankind, will be derived from the manufactory and general use of maple sugar, for the benefits which I flatter myself are to result from it, will not be confined to our own country. They will, I hope, extend themselves to the interests of humanity in the West-Indies. With this view of the subject of this letter, I cannot help contemplating a sugar maple tree with a species of affection and even veneration, for I have persuaded myself to behold in it the happy means of rendering the commerce and slavery of our African brethren, in the sugar Islands as unnecessary, as it has always been inhuman and unjust.

*Source*: Benjamin Rush to Thomas Jefferson, July 10, 1791, in Benjamin Rush, *Essays, Literary, Moral and Philosophical* (Philadelphia: Bradford, 1806), pp. 270, 282, 284–85, 287.

## DOCUMENT 25: The Founding Fathers on the Care of the Land (1793, 1818)

Three of the first four U.S. presidents—George Washington, Thomas Jefferson, and James Madison—were Virginia planters who devoted a great deal of thought, study, and effort to the development of techniques for making their land more productive. While they carefully rotated crops and fields, they rejected the use of high-priced fertilizers until rising land prices made such expenditures practical. They were all active members of the American Philosophical Society, one of whose primary goals was the furtherance of scientific agriculture.

### A. Thomas Jefferson to George Washington, June 28, 1793

Manure does not enter into this [the making of a good farm] because we can buy an acre of new land cheaper than we can manure an old acre.

### B. Thomas Jefferson to Thomas Mann Rudolphe, July 28, 1793

[Dr. George Logan] thinks that the whole improvement in the modern agriculture of Europe consists in the substitution of red clover instead of 3. years of fallow or rest, whether successive or interspersed leaves the land much heartier at the close of the rotation; that there is no doubt of this fact, the difference being palpable.

### C. George Washington to Thomas Jefferson, n.d.

I permit no separate inclosures of my fields. their limits are preserved by 2. Rows of peach trees, leaving a road between them. my fields are by this means protected from pasturage as follows

Wheat after 2. years of clover, the clover turned in in autumn by [illegible] ploughing, the wheat sowed on that & buried by a harrow drawn the direction of the furrows. . . . as soon as the wheat is cut I propose (as soon as I can get the winter vetch) to turn in the stubble, sow vetch and cut it for green fodder in Feb. & March. Then turn in the stubble of that as a green . . . dressing, & the ground is ready for [illegible] in alternate rows 4½ feet apart. Put into the drills the long dung which has been made from the straw of this field in the preceding winter, in autumn sow vetch again.

### D. James Madison on Intelligent Husbandry, 1818

The error first to be noticed [in our husbandry] is that of cultivating land, either naturally poor or impoverished by cultivation. This error, like many others, is the effect of habit, continued after the reason for it has failed. Whilst there was an abundance of fresh and fertile soil, it was the interest of the cultivator to spread his labor over as great a surface as he could. Land being cheap and labor dear, and the land co-operating powerfully with the labor, it was profitable to draw as much as possible from the land. Labor is now comparatively cheaper and land dearer. . . . It might be profitable, therefore, now to contract the surface over which labor is spread, even if the soil retained its freshness and fertility. But this is not the case. Much of the fertile soil is exhausted, and unfertile soils are brought into cultivation.

. . .

The evil of pressing too hard upon the land, has also been much increased by the bad mode of ploughing it. Shallow ploughing, and ploughing up and down hilly land, have, by exposing the loosened soil to be carried off by rains, hastened more than any thing else, the waste of its fertility.

. . .

The neglect of manures is another error which claims particular notice. It may be traced to the same cause with our excessive cropping. In the early stages of our agriculture, it was more convenient, and more profitable, to bring new land into cultivation, than to improve exhausted land.

*Source*: **A–C**. Edwin Morris Betts, ed., *Thomas Jefferson's Farm Book* (Princeton, NJ: Princeton University Press, 1953), pp. 188–89, 194, 314. **D**. "Agricultural Society of Albemarle; Address by Mr. Madison," *Niles' Weekly Register*, Vol. II (New Series), No. 21 (July 18, 1818), in Robert McHenry and Charles Van Doren, eds., *A Documentary History of Conservation in America* (New York: Praeger, 1972), pp. 273–75.

---

## DOCUMENT 26: Thomas Malthus's *Essay on the Principle of Population* (1798)

By the seventeenth century, population pressures in Europe had created an increasing need for farmland and a demand for changes in landownership patterns. Thomas Malthus's apocalyptic essay theorizing that population, if unchecked, multiplies geometrically while the food supply multiplies only arithmetically and that this will eventually result in a food crisis was an outgrowth of an ongoing discussion among economists in France and England about how to deal with population growth. At the time Malthus published his essay, many Americans, including James Madison [see Document 21], were familiar with these discussions but did not consider them to have much bearing on the contemporary U.S. situation. Two and a half decades later, in a letter to Edward Everett (a Harvard professor and governor of Massachusetts), Madison complained that Malthus had ignored the effects of political and social conditions on population growth. Indeed, by the end of the nineteenth century, the growth of the U.S. population, the increase in poverty and the closing of the frontier had made food supply and land availability relevant issues on this side of the Atlantic, and in the mid-twentieth century doomsayers, such as Paul Ehrlich [see Document 106], could be heard repeating the Malthusian arguments, including his law of "diminishing returns," which states that, as time goes by, increasingly greater effort is required to obtain the same yield from a mine, a forest, or a piece of land.

## A. Malthus's Essay

I think I may fairly make two postulata.

First, That food is necessary to the existence of man.

Secondly, That the passion between the sexes is necessary and will remain nearly in its present state.

These two laws, ever since we have had any knowledge of mankind, appear to have been fixed laws of our nature, and, as we have not hitherto seen any alteration in them, we have no right to conclude that they will ever cease to be what they now are, without an immediate act of power in that Being who first arranged the system of the universe, and for the advantage of his creatures, still executes, according to fixed laws, all its various operations.

I do not know that any writer has supposed that on this earth man will ultimately be able to live without food. But Mr. [William] Godwin has conjectured that the passion between the sexes may in time be extinguished. As, however, he calls this part of his work a deviation into the land of conjecture, I will not dwell longer upon it at present than to say that the best arguments for the perfectibility of man are drawn from a contemplation of the great progress that he has already made from the savage state and the difficulty of saying where he is to stop. But towards the extinction of the passion between the sexes, no progress whatever has hitherto been made. It appears to exist in as much force at present as it did two thousand or four thousand years ago. . . .

Assuming then, my postulata as granted, I say that the power of population is indefinitely greater than the power in the earth to produce subsistence for man.

Population, when unchecked, increases in a geometrical ratio. Subsistence increases only in an arithmetical ratio. A slight acquaintance with numbers will shew the immensity of the first power in comparison of the second.

By that law of our nature which makes food necessary to the life of man, the effects of these two unequal powers must be kept equal.

This implies a strong and constantly operating check on population from the difficulty of subsistence. This difficulty must fall some where and must necessarily be severely felt by a large portion of mankind.

Through the animal and vegetable kingdoms, nature has scattered the seeds of life abroad with the most profuse and liberal hand. She has been comparatively sparing in the room and the nourishment necessary to rear them. The germs of existence contained in this spot of earth, with ample food and ample room to expand in, would fill millions of worlds in the course of a few thousand years. Necessity, that imperious all pervading law of nature, restrains them within the prescribed bounds. The race of plants and the race of animals shrink under this great restrictive law. And the race of man cannot, by any efforts of reason, escape from it. Among plants and animals its effects are waste of seed, sickness, and premature death. Among mankind, misery and vice.

### B. James Madison to Edward Everett, November 26, 1823

That the rate of increase in the population of the U.S. is influenced at the same time by their political & social condition is proved by the slower increase under the vicious institutions of Spanish America where Nature was not less bountiful. Nor can it be doubted that the actual population of Europe w$^d$ be augmented by such reforms in the systems as would enlighten & animate the efforts to render the funds of subsistence more productive. We see everywhere in that quarter of the Globe, the people increasing in number as the ancient burdens & abuses have yielded to the progress of light & civilization. . . .

Mr. Malthus has certainly shewn much ability in his illustrations & applications of the principle he assumes, however much he may have erred in some of his positions. But he has not all the merit of originality which has been allowed him. The principle was adverted to & reasoned upon, long before him, tho' with views & applications not the same with his. The principle is indeed inherent in all the organized beings on the Globe, as well of the animal as the vegetable classes; all & each of which when left to themselves, multiply till checked by the limited fund of their pabulum, or by the mortality generated by an excess of their numbers.

*Source*: **A.** Thomas Robert Malthus, *An Essay on the Principle of Population*, ed. Philip Appleman (New York: W. W. Norton, 1976), pp. 18, 19–20. **B.** Saul K.

Padover, ed., *The Complete Madison: His Basic Writings* (New York: Harper, 1953), p. 322.

## DOCUMENT 27: Meriwether Lewis on the Slaughter of Buffaloes (1805)

In their expedition to find a practical route to the Pacific Ocean (the fabled Northwest Passage) and, simultaneously, to explore the new lands acquired in the Louisiana Purchase, Captain Meriwether Lewis and William Clark traveled up the Missouri River to its source, across the Great Divide in the Rocky Mountains, and then down the Columbia River to the Pacific Ocean. The purposes of this first official U.S. expedition and first official survey of some of the nation's resources were commercial, political, and scientific—to acquire data about the geographic features of the land, Indian life and culture, and the flora and fauna west of the Missouri. One of Lewis's conclusions was that much of the land through which they had passed was unsuitable for intensive farming.

Lewis took note in his journals of the great diversity of Indian cultures and commented on variations in diet, housing, and farming, fishing, and hunting techniques. In this selection, Lewis describes an Indian method of hunting bison without using a horse, gun, or even a bow and arrow. Use of the buffalo jump technique, which enabled Indians to kill 100 to 200 animals at a time, had little impact on the bison population compared with the massive slaughters that the whites were able to carry out using firearms, horses, and horse-drawn vehicles.

today we passed on the Star^d. side the remains of a vast many mangled carcases of Buffalow which had been driven over a precipice of 120 feet by the Indians and perished; the water appeared to have washed away a part of this immence pile of slaughter and still their remained the fragments of at least a hundred carcases they created a most horrid stench. in this manner the Indians of the Missouri distroy vast herds of buffaloe at a stroke; for this purpose one of the most active and fleet young men is scelected and disguised in a robe of buffaloe skin, having also the skin of the buffaloe's head with the years and horns fastened on his head in form of a cap, thus caparisoned he places himself at a convenient distance between a herd of buffaloe and a precipice proper for the purpose, which happens in many places on this river for miles together; the other indians now surround the herd on the back and flanks and at a signal agreed on all shew themselves at the same time moving forward towards the buffaloe; the disguised indian or decoy has taken care to place himself sufficiently nigh the buffaloe to be noticed by them when they take to flight and runing before them they follow him in full speede to the precipice, the cattle behind driving those in front over and seeing them go do not look or hesitate about following untill the whole are precipitated down the precipice forming one common mass of dead an[d] mangled carcases: the decoy in the mean time has taken care to secure himself in some cranney or crivice of the clift which he had previously prepared for that purpose. the part of the decoy I am informed is extremely dangerous, if they are not very fleet runers the buffaloe tread them under foot and crush them to death, and sometimes drive them over the precipice also, where they perish in common with the buffaloe. we saw a great many wolves in the neighbourhood of these mangled carcases they were fat and extreemly gentle.

*Source*: Meriwether Lewis, Journal entry for May 29, 1805, in Reuben Gold Thwaites, *Original Journals of the Lewis and Clark Expedition, 1804–1806*, Vol. 2 (New York: Antiquarian Press, 1959; reprint of edition of 1905), pp. 93–94.

## DOCUMENT 28: Act Establishing the First Federal Forest Reserve (1817)

The first national effort to set aside forest lands was undertaken during the administration of James Madison. It was a pragmatic act to ensure the fledgling U.S. Navy of adequate timber supplies for shipbuilding. Madison's secretary of state, John Quincy Adams, who as president would later expand the nation's forest reserves, was undoubtedly a supporter of the act.

*Be it enacted . . .* That the Secretary of the Navy be authorized, and it shall be his duty, under the direction of the President of the United States, to cause such vacant and unappropriated lands of the United States as produce the live oak and red cedar timbers to be explored, and selection to be made of such tracts or portions thereof, where the principal growth is of either of the said timbers, as in his judgment may be necessary to furnish for the navy a sufficient supply of the said timbers. The said Secretary shall have power to employ such agent or agents and surveyor as he may deem necessary for the aforesaid purpose, who shall report to him the tracts by them selected, with the boundaries ascertained and accurately designated by actual survey or water courses, which report shall be laid before the President, which he may approve or reject in whole or in part; and the tracts of land thus selected with the approbation of the President, shall be reserved unless otherwise directed by law, from any future sale of the public lands, and be appropriated to the sole purpose of supplying timber for the navy of the United States.

*Source*: George P. Sanger, ed., *The Statutes at Large, Treaties, and Proclamations of the United States of America*, 14th Cong., 2nd sess., chap. 22, March 1, 1817, p. 347.

---

## DOCUMENT 29: Act to Protect Useful Birds in Massachusetts (1818)

The earliest law to protect birds in the United States was an act by the Massachusetts legislature designated "An Act to prevent the destruction of certain useful Birds at unseasonable times of the year." It was probably the first law to recognize nature's services.

Whereas there are within the Commonwealth, many birds which are useful and profitable to the citizens, either as articles of food, or as instruments in the hands of Providence to destroy various noxious insects, grubs and caterpillars, which are prejudicial or destructive to vegetation, fruits and grain; and it is desirable to promote the increase and preservation of birds of the above description, and to prevent the wanton destruction of them at improper seasons:

... hereafter it shall not be lawful for any person to take, kill or destroy, any of the birds called partridges and quails, at any time from the first day of March, to the first day of September in every year; and no person shall take, kill or destroy, any of the birds called woodcocks, snipes, larks and robins, at any time from the first day of March to the fourth day of July in each year; and if any person shall take or kill, or shall sell, buy or have in his possession after being killed, or taken, any of the birds aforesaid, within the times limited as aforesaid respectively, he shall forfeit and pay for each and every partridge, quail or woodcock, so taken, killed or in his possession, two dollars; and for each and every snipe, lark or robin, so killed, taken, or in his posession, one dollar.

*Source: Laws of the Commonwealth of Massachusetts*, Vol. 7, chap. 103 (Boston: 1818), in Robert McHenry and Charles Van Doren, eds., *A Documentary History of Conservation in America* (New York: Praeger, 1972), pp. 272–73.

---

## DOCUMENT 30: James Fenimore Cooper Laments the Disappearance of Unregulated Wilderness (1823)

The woodsman Natty Bumppo (also called Leatherstocking), the main character in the James Fenimore Cooper Leatherstocking series, here laments the loss of open land in New York State. He holds that wild game should belong to whoever kills it, while the judge claims that it belongs to the owner of the property on which it was killed. The difference in viewpoints, as Madison pointed out [see Document 21B], was produced by the difference in the property holdings of the two claimants.

"The legislature have been passing laws," continued [Judge] Marmaduke [Temple], "that the country much required. Among others, there is an act, prohibiting the drawing of seines, at any other than proper seasons, in certain of our streams and small lakes; and another, to prohibit the killing of deer in the teeming months. These are laws that were loudly called for, by judicious men; nor do I despair of getting an act, to make the unlawful felling of timber a criminal offence."

The hunter [Natty Bumppo] listened to this detail with breathless attention, and when the Judge had ended, he laughed in open derision for a moment, before he made this reply:—

"You may make your laws, Judge, but who will you find to watch the mountains through the long summer days, or the lakes at night? Game is game, and he who finds may kill; that has been the law in these mountains for forty years, to my sartain knowledge; and I think one old law is worth two new ones. None but a green-one would wish to kill a doe with a fa'n by its side, unless his moccasins was getting old, or his leggins ragged, for the flesh is lean and coarse. But a rifle rings along them rocks along the lake shore, sometimes, as if fifty pieces were fired at once:—it would be hard to tell where the man stood who pulled the trigger."

"Armed with the dignity of the law, Mr. Bumppo," returned the Judge, gravely, "a vigilant magistrate can prevent much of the evil that has hitherto prevailed, and which is already rendering the game scarce. I hope to live to see the day when a man's rights in his game shall be as much respected as his title to his farm."

"Your titles and your farms are all new together," cried Natty; "but laws should be equal, and not more for one than another. I shot a deer, last Wednesday was a fortnight, and it floundered through the snowbanks till it got over a brush fence; I catch'd the lock of my rifle in the twigs, in following, and was kept back, until finally the creater got off. Now I want to know who is to pay me for that deer; and a fine buck it was. If there hadn't been a fence, I should have got another shot into it. . . .—No, no, Judge, it's the farmers that makes the game scearce, and not the hunters."

"Ter teer is not so plenty as in ter old war, Pumppo," said the Major, who had been an attentive listener, amidst clouds of smoke; "put ter lant is not mate for ter teer to live on, put for Christians."

"Why, Major, I believe you're a friend to justice and the right, though you go so often to the grand house [Temple's house]; but it's a hard case to a man to have his honest calling for a livelihood stopt by sitch laws, and that too when, if right was done, he mought hunt or fish on any day in the week, or on the best flat in the Patent, if he was so minded."

"I unterstant you, Letter-stockint," returned the Major, fixing his black eyes, with a look of peculiar meaning, on the hunter; "put you tidn't use to be so prutent, as to look ahet mit so much care."

"Maybe there wasn't so much 'casion," said the hunter, a little sulkily; when he sunk into a profound silence, from which he was not roused for some time.

*Source*: James Fenimore Cooper, *The Pioneers* (London: Allman, 1823), pp. 150–51.

## DOCUMENT 31: George Catlin's Proposal for a National Park (1832)

George Catlin, who traveled in the West from 1832 to 1839, gathering information about American Indians and painting their portraits in natural settings, was the first American to recognize that, without government protection, the western wilderness would be overrun by settlers. Unfortunately, the grasslands of the Great Plains did not become objects of government conservation policy until after they been had plowed under.

Many are the rudenesses and wilds in Nature's works, which are destined to fall before the deadly axe and desolating hands of cultivating man; and so amongst her ranks of *living*, of beast and human, we often find noble stamps, or beautiful colours, to which our admiration clings; and even in the overwhelming march of civilised improvements and refinements do we love to cherish their existence, and lend our efforts to preserve them in their primitive rudeness. Such of Nature's works are always worthy of our preservation and protection; and the further we become separated (and the face of the country) from pristine wildness and beauty, the more pleasure does the mind of enlightened man feel in recurring to those scenes, when he can have them preserved for his eyes and his mind to dwell upon.

Of such "rudenesses and wilds," Nature has nowhere presented more beautiful and lovely scenes, than those of the vast prairies of the West; and of *man* and *beast*, no nobler specimens than those who inhabit them—the *Indian* and the *buffalo*—joint and original tenants of the soil, and fugitives together from the approach of civilized man.

* * *

This strip of country, which extends from the province of Mexico to Lake Winnipeg on the North, is almost one entire plain of grass, which is, and ever must be, useless to cultivating man. It is here, and here chiefly, that the buffaloes dwell; and with, and hovering about them, live and flourish the tribes of Indians, whom God made for the enjoyment of that fair land and its luxuries.

It is a melancholy contemplation for one who has travelled as I have, through these realms, and seen this noble animal in all its pride and glory, to contemplate it so rapidly wasting from the world, drawing the irresistible conclusion too, which one must do, that its species is soon to be extinguished, and with it the peace and happiness (if not the actual existence) of the tribes of Indians who are joint tenants with them, in the occupancy of these vast and idle plains.

And what a splendid contemplation too, when one (who has travelled these realms, and can duly appreciate them) imagines them as they *might* in future be seen (by some great protecting policy of government) preserved in their pristine beauty and wildness, in a *magnificent park*, where the world could see for ages to come, the native Indian in his classic attire, galloping his wild horse, with sinewy bow, and shield and lance, amid the fleeting herds of elks and buffaloes. What a beautiful and thrilling specimen for America to preserve and hold up to the view of her refined citizens and the world, in future ages! A *nation's Park*, containing man and beast, in all the wild and freshness of their nature's beauty!

I would ask no other monument to my memory, nor any other enrolment of my name amongst the famous dead, than the reputation of having been the founder of such an institution.

*Source*: George Catlin, letter to *New York Commercial Advertiser*, in Catlin, *North American Indians: Being Letters and Notes on Their Manners, Customs, and Conditions, Written during Eight Years' Travel amongst the Wildest Tribes in North America, 1832–39*, Vol. 1 (London, 1880), pp. 294–95.

---

## DOCUMENT 32: Black Hawk on the Indians and the Land (1833)

The Sac Indian chief Black Hawk led the Sac and Fox in battle against the whites who came to take their farms. In his memoirs, he described the rich Illinois land that his tribe lost and commented on its place in the lives of his people.

Our village was situated on the north side of Rock River, at the foot of the rapids, on the point of land between Rock River and the Mississippi. . . .

On its highest peak our Watch Tower was situated, from which we had a fine view for many miles up and down Rock River, and in every direction. On the side of this bluff we had our cornfields, extending about two miles up parallel with the larger river, where they adjoined those of the Foxes, whose village was on the same stream, opposite the lower end of Rock Island, and three miles distant from ours. We had eight hundred acres in cultivation including what we had on the islands in Rock River. The land around our village which remained unbroken was covered with blue grass which furnished excellent pasture for our horses. Several fine springs poured out of the bluff near by, from which we were well supplied with good water. The rapids of Rock River furnished us with an abundance of excellent fish, and the land being very fertile, never failed to produce good crops of corn, beans, pumpkins, and squashes. We always had plenty; our children never cried from hunger, neither were our people in want. Here our village had stood for more than a hundred years, during all of which time we were the undisputed possessors of the Mississippi Valley, from the Wisconsin to the Portage des Sioux, near the mouth of the Missouri, being about seven hundred miles in length.

At this time we had very little intercourse with the whites except those who were traders. Our village was healthy, and there was no place in the country possessing such advantages, nor hunting grounds better than those we had in possession. If a prophet had come to our village in those days and told us that the things were to take place which have since come to pass, none of our people would have believed him. What! to be driven from our village and our hunting grounds, and not even to be permitted to visit the graves of our forefathers and relatives and friends?

\* \* \*

My reason teaches me that land cannot be sold. The Great Spirit gave it to his children to live upon, and cultivate, as far as necessary for their subsistence; and so long as they occupy and cultivate it they have the right to the soil, but if they voluntarily leave it then any other people have a right to settle upon it. Nothing can be sold but such things as can be carried away.

In consequence of the improvements of the intruders on our fields, we found considerable difficulty to get ground to plant a little corn Some of the whites permitted us to plant small patches [of corn] in the fields they had fenced, keeping all the best ground for themselves. Our women had great difficulty in climbing their fences, being unaccustomed to the kind, and were ill-treated if they left a rail down.

One of my old friends thought he was safe. His cornfield was on a small island in Rock River. He planted his corn, it came up well; but the white man saw it, he wanted it [the island], and took his team over, ploughed up the corn, and replanted it for himself. The old man shed tears; not for himself, but on account of the distress his family would be in if they raised no corn.

*Source: Black Hawk's Autobiography*, interpreted by Antoine LeClaire, ed. J. B. Patterson and James D. Rishell (Rock Island, IL: American Publishing, 1912), pp. 62–63, 84–85.

---

## DOCUMENT 33: John James Audubon on the Senseless Destruction of Fish, Birds, and Quadrupeds (1833)

The Haitian-born naturalist and painter John James Audubon traveled throughout the United States and parts of Canada in search of subjects for his paintings of the birds and quadrupeds of North America. He deplored the senseless destruction of America's wildlife that was obviously taking place all around him.

We are often told rum kills the Indian; I think not; it is oftener the want of food, the loss of hope as he loses sight of all that was once abundant, before the white man intruded on his land and killed off the wild quadrupeds and birds with which he has fed and clothed himself since his creation. Nature herself seems perishing. Labrador must shortly be depeopled, not only of aboriginal man, but of all else having life, owing to man's cupidity. When no more fish, no more game, no more birds exist on her hills, along her coasts, and in her rivers, then she will be abandoned and deserted like a worn-out field.

*Source*: John James Audubon, *Labrador Journals*, in A. Donald Culross Peattie, ed., *Audubon's America* (Boston: Houghton Mifflin, 1940), p. 245.

# Part III

# The Origins of Environmental Activism, 1840–1890

The acquisition of the Oregon Territory in 1846 and the Mexican Cession of 1848 expanded the United States' western border to the Pacific Coast, while the annexation of Texas in 1845 and the Gadsden purchase of 1853 together with the Mexican Cession defined the country's southern border. By 1867, with the purchase of Alaska from Russia, all the land that was to form the continental United States had become part of the nation.

## INDUSTRIAL AND URBAN GROWTH

While the country was in the throes of its great westward expansion, the landscape of the eastern portion of the country was undergoing stupendous change. In 1776 New York and Philadelphia were the only cities with more than 20,000 inhabitants; by 1860 there were forty-three metropolises with populations of at least that size. Not only was the number of cities growing, but the size of those cities was also expanding rapidly. The population of New York City, for example, was estimated to be 75,770 in 1805; by 1850 it had risen to 515,000, and by 1890, in Manhattan alone, it had reached 1,441,216. The 1890 population for the whole city (which by then included Brooklyn) was over 2.5 million.

The growth of the urban population between 1840 and 1890 was in great part the result of a tremendous flood of immigrants from western Europe that carried nearly 10 million newcomers to America's shores. It also reflected the nation's accelerating industrialization and shifting pattern of employment. In 1861 there were nearly 31.5 million people in the United States, 19.8 percent of whom lived in large towns and cities and fewer than half of whom were farmworkers. By 1880 the population had swelled to more than 50 million people, 28.2 percent of them living in urban areas.

During the Civil War (1861–1865), which pitted the increasingly urban North against the predominantly agrarian South, the need for armaments and other wartime goods spurred the expansion of industry and brought workers into towns and cities to work in the factories. Especially in the larger cities along the eastern seaboard, the vast majority of new urban residents and industrial workers came from among the tens of thousands of immigrants pouring into the United States every year.

## URBAN SANITATION AND INFRASTRUCTURE PROBLEMS

Industrial and urban expansion were accompanied by a host of problems: smoke and soot, noise, garbage, poor sanitation and sewage, poor drainage, inadequate and unclean water supplies, and droppings from horses [see Documents 40, 41, and 46]. Doctors and others concerned about human health and well-being pressed for improved garbage collection and sanitation, regulated quarantines during epidemics, more widespread vaccination against smallpox, and the collection of accurate data concerning births and deaths. Their efforts marked the beginning of the sanitary movement, which led to the formation of public health regulatory organizations and raised the consciousness of both the public and local governments about the need for better sanitation and health record keeping.

At midcentury few cities other than Boston, New York, and Philadelphia had public sewage systems. (Boston's system actually dated from the seventeenth century.) Except in these few cities, indoor plumbing was to be found mainly in the homes of the wealthy. In cities and towns that lacked sewer systems, drain pipes from indoor plumbing emptied into pits beneath people's homes or in their yards. Those who were without indoor plumbing also emptied their chamber pots and wastewater pans into pits in their yards or beneath their outhouses. By the 1850s, though, municipalities with booming populations, such as Brooklyn and Chicago, were forced to draw up plans for sewer systems, both to eliminate with the stench emanating from the slop holes and for health reasons.[1] These new sewer systems, however, like the older sewer systems in Boston and New York, were designed simply to dump raw sewage into nearby rivers, streams, or harbors.

Rapidly increasing populations greatly strained city water supplies, and the growth in industrialization placed added demands on municipal water supply systems. Prior to the 1840s, most city residents depended on local communal wells, cisterns, springs, streams, and transported water for their water, but by the 1840s local water supplies were beginning to prove inadequate. Even municipalities that for years had supplemented their well water with local reservoirs found it necessary to draw on more distant water sources. In 1842, for example, New York City built the Croton Aqueduct to carry water from a reservoir in Westchester County, forty-one miles north of the city.[2] As indoor running water and flush toilets became commonplace in urban areas and as the nation became more industrial, the per capita usage of water increased, placing ever greater demands on municipal water supplies.

Keeping the water supply clean proved very difficult. During periods of heavy rain, there was constant danger of contamination of municipal wells by runoff from the slop pits and manure-laden streets. Microbial diseases such as typhoid fever were quickly spread when wells became contaminated.

## RAILROADS AND WESTWARD EXPANSION

One of the major spurs to industrial development was the spread of the railroads. The completion of the New York–Chicago rail link in 1853 and the first transcontinental rail route in 1869 gave impetus not only to industrialization but also to resource exploitation and the settlement of the West. People and manufactured goods could be transported across the country quickly and easily, and timber, coal, iron, and, later, oil could be carried to mills, factories, and refineries. The discovery of iron near Lake Superior in 1844 and of gold in Sutter's Mill, California, in 1848 and at Pikes Peak, Colorado, in 1859 spurred the westward movement of settlers and entrepreneurs.

To encourage the settlement of the West, Congress passed legislation that enabled settlers to buy land cheaply [see Document 42] and have easy access to land with mineral deposits [see Document 51] and grazing lands. Low purchase prices and—if the lands were to remain in the public domain—leasing fees became the norm. Vast tracts of land were turned over to the railroad companies as inducements for building additional railroads, and the railroad companies advertised in both Europe and the eastern part of the United States for new settlers to make the westward journey. The expansion of railroads, mining operations, grain fields, and factories was financed by bank loans; when this growth proved too rapid for the country to absorb, many banks failed, resulting in the panic of 1873.

The panic prompted some national soul searching and a questioning of federal policy. Reformers, including the social economist Henry George, attacked national land policies as enormous giveaways designed to enrich speculators and powerful real estate, industrial, mining, and ranching interests. In place of prevailing practices, George advocated a national land development policy that would take into account the public well-being and the interests of ordinary people [see Document 53].

Although the growth of the railroads had proved a boon to the expansion of the country and the development of industry, it posed an unforeseen threat to the nation's health. The danger first became evident during the yellow fever epidemic of 1879, which began in Memphis and quickly spread to nearby states as already-infected people traveled by rail to escape the plague. A National Board of Health was formed by Congress to help prevent the further spread of the disease and to study sanitation in Memphis.[3]

## SCIENTIFIC INNOVATION

During the late nineteenth century enormous changes were taking place in science as well as industry. Building on a foundation laid down by John Ray [see Document 15] and Carl Linnaeus in the first half of the eighteenth century, naturalists working in both Europe and America in the late eighteenth and early nineteenth centuries had attempted to identify, describe, and classify all the known species of plants and animals. But these natural historians accepted the idea that each species was created by God and had retained a specific, immutable structure from the beginning of time. By the middle of the nineteenth century, however, scores of botanists, zoologists, and natural philosophers in western and central Europe and in the United States, buoyed by a liberal political and social atmosphere, were occupied with comprehending how changes in species occurred and were attempting to understand the relationships among organisms and between organisms and their environments.

In England, in 1858, the Linnean Society of London published papers by Alfred Russel Wallace and Charles Darwin on the theory of natural selection. The two men had independently developed the same theory—Wallace, based on his eight years of travel through the Malay archipelago, and Darwin, based on his around-the-globe voyage on the H.M.S *Beagle*, during which he spent more than four years doing studies along the coast and the nearby interior of South America, sailed through the Galapagos archipelago, and visited Tahiti and New Zealand. A year after presenting his theory to the Linnean Society, Darwin published *The Origin of Species*, a book that produced controversy on both sides of the Atlantic and sparked the imagination of the public as well as the scientific community. In the United States, in 1864, George Perkins Marsh published his monumental *Man and Nature* [see Document 43], the first wide-ranging, scholarly study of how human actions affect the natural world around them. In Moravia in 1865, Gregor Mendel wrote his laws concerning the inheritance of physical characteristics (but which remained unknown outside Moravia until the twentieth century), and in Germany in 1866, Ernst Haeckel published his *Generelle Morphologie der Organismen*, in which he coined the term *ecology*, defining it as "the comprehensive science of the relationships of the organism to the environment."[4] Then, in 1871, Darwin published *The Descent of Man* [see Document 49], which theorized that humans are part of the natural system and not a unique creation, separate from all other living things. The idea that humans might have evolved from a common ancestry with apes caused an enormous public uproar, whose echoes can still be heard.

The studies on which these writings were based constituted the beginning of modern scientific investigation and were the hallmarks of an era when scientists and talented amateurs formed a large number of organizations to promote scientific study. While most of these groups—such as the American Ornithologists' Union (AOU), organized in 1883—had a very specific scientific focus, the most prominent of the American societies, the American Association for the Advancement of Science (AAAS), founded in 1848, attempted to reach out to a wide range of professional scientists and to establish a recognized forum for them. Many of the American scientific societies established in the middle of the nineteenth century, including the AAAS and the AOU, continue to play an important role in the furtherance of scientific research today.

## THE NEED FOR CONTACT WITH NATURE

As the eastern part of the nation became increasingly urban and industrial, as technology more and more limited people's contact with nature, and as awareness of the natural beauty of the western landscape increasingly filtered into the national consciousness, American writers and artists began to idealize the natural world. The romantic movement, which developed in Europe in the latter part of the eighteenth century partly as a reaction to the Industrial Revolution and its distancing of people from nature, influenced the depiction of the wilderness by both the Hudson River school painters [see Document 34] and the American transcendentalists [see Documents 36 and 44].

Henry David Thoreau [Document 44], John Muir [Document 47], and Frederick Law Olmsted and Calvert Vaux [Document 48] were among the ardent and eloquent advocates of the need for people to get away from urban, industrialized areas and the importance of setting aside space for this purpose. A recognition of the human need for nature as well as a sense that open space was disappearing inspired foresighted individuals to suggest that land be set aside for parks. Among the first of these were the painter George Catlin [see Document 32], the poet and journalist William Cullen Bryant [see Document 37], and the journalist and landscape architect Andrew Jackson Downing [see Document 38]. Bryant and Downing, in planting the idea for a large-scale public park in New York City, launched the urban parks movement, the first concerted conservation effort in the United States. Cities, states, and the nation as a whole began to recognize the need to set aside land for human recreation and spiritual restoration, as well as to preserve places of unique natural beauty. Before the end of the century, a host of magnificent urban, state, and national parks were created, including Central and Prospect parks in New York City [see Document 48], Fairmont Park in Philadelphia, Yosemite Park (originally chartered as a California state park but later to become a national park) [see Document 45], New York State's Adirondack Forest Preserve [see Document 54], and Yellowstone, the first national park [see Document 50].

## THE BEGINNING OF THE CONSERVATION IDEA

As the new century approached, increasing numbers of people—primarily wealthy individuals—became aware that not only were America's wilderness areas beginning to disappear, but that the wildlife that inhabited them was also being destroyed. In 1865, 15 million bison roamed the Great Plains; by 1885 only about 3,000 bison remained in the United States.

Organized resistance to the wanton slaughter of birds and mammals and to the destruction of habitats suitable for fish and game began to coalesce around individuals, such as George Bird Grinnell and Theodore Roosevelt, who realized that our resources were disappearing and who had the journalistic or political clout to effect change. Grinnell started the Audubon Society in 1886, with one of its goals being to stop the killing of birds to provide feathers for women's hats [see Document 56]. Scientific organizations also actively lobbied at both the state and federal levels for the passage of laws that encouraged conservation. In 1873 the AAAS, in a memo to Congress, requested that laws be passed to protect the nation's natural resources, and Grinnell's bird protection drive was aided by the American Ornithologists' Union, which prepared a model bird protection law [see Document 55] and printed and distributed 100,000 copies of it.

Other conservation-minded groups chartered in the last three decades of the nineteenth century included the Appalachian Mountain Club, formed in 1876, and the Sierra Club, organized in 1892 by John Muir—both of which were primarily hiking clubs that advocated the preservation of scenic lands for the enjoyment of the public—and the

Boone and Crockett Club, established in 1887 [see Document 57], which was primarily a hunting club that advocated the conservation of the habitats of game animals. These use-oriented organizations, together with the scientific societies and the advocacy-oriented state Audubon societies, formed a base of organized public support for the nascent conservation movement.

While these nongovernmental organizations were becoming established, the federal government was beginning to look at issues related to the management and conservation of resources. The government's main concerns were the management of timber, water, minerals, and land. Toward the end of the nineteenth century a number of people were brought into government service to manage resources and help develop resource management policies. The recommendations and actions of three of these individuals—Carl Schurz, Gifford Pinchot, and John Wesley Powell—greatly affected the course of U.S environmental policy for more than half a century.

Schurz, who served as secretary of the Interior from 1877 to 1881, promoted the scientific management of forests and recommended that timberland be set aside [see Document 52]. Pinchot, the first professional American forester, served as head of the U.S. Forestry Service. Powell joined the staff of the U.S. Geological Survey in 1875 and served as its head from 1881 to 1894. After touring the Southwest, Powell began a campaign to "reclaim" the arid lands of that region [see Document 58].

Powell's campaign paralleled the efforts of Louisianans and Alabamans to convince the federal government to enact legislation that would allow them to drain wetlands [see Document 39]—swamp and overflow land not suited for cultivation—so that they could sell the "reclaimed" land to land-hungry buyers. As a result of these efforts, the large-scale manipulation of the environment to develop agricultural lands, water resources, and hydroelectric power had become established federal policy in both arid regions and wetlands by the beginning of the twentieth century. The environmentally disastrous consequences of this policy would not be widely visible for another half-century.

## Document 34: Thomas Cole's "Lament of the Forest" (1841)

A distinguished member of the Hudson River school of landscape painters, Thomas Cole, along with other adherents of the romantic movement, viewed the spread of factories across the land not only as a blight on the natural landscape but also as an encroachment on the human spirit. Cole and other Hudson River school artists, such as Asher B. Durand, painted pictures of well-dressed people walking in rustic settings (e.g., Durand's *Kindred Spirits*, painted in 1849, depicts William Cullen Bryant and Thomas Cole in a woodland scene), and these pictures helped to make excursions to rustic areas fashionable.

> ... Our doom is near; behold from east to west
> The skies are darkened by ascending smoke;
> Each hill and every valley is become
> An altar unto Mammon, and the gods
> Of man's idolatry—the victims we.
> Missouri's floods are ruffled as by storm,
> And Hudson's rugged hills at midnight glow
> By light of man-projected meteors.
> We feed ten thousand fires: in our short day
> The woodland growth of centuries is consumed. . . .

* * *

> A few short years!—these valleys, greenly clad,
> These slumbering mountains, resting in our arms,
> Shall naked glare beneath the scorching sun,
> And all their wimpling rivulets be dry.
> No more the deer shall haunt these bosky glens,
> Nor the pert squirrel chatter near his store.

*Source*: Thomas Cole, "Lament of the Forest," *Knickerbocker*, 17, no. 6 (June 1841), in Robert McHenry and Charles Van Doren, eds., *A Documentary History of Conservation in the America* (New York: Praeger, 1972), p. 175.

---

## DOCUMENT 35: John James Audubon on the Decimation of the Bison Herds (1843)

---

Audubon was one of the first Americans to call for an end to the wanton slaughter of bison.

July 21

... we could see [the buffalo chase] when nearly a mile distant.
. . .
What a terrible destruction of life, as it were for nothing, or next to it, as the tongues only were brought in, and the flesh of these fine animals

was left to beasts and birds of prey, or to rot on the spots where they fell. The prairies are literally *covered* with the skulls of the victims, and the roads the Buffalo make in crossing the prairies have all the appearance of heavy wagon tracks.

## August 2

Buffaloes become so very poor during hard winters, when the snows cover the ground to the depth of two or three feet, that they lose their hair, become covered with scabs, on which the Magpies feed, and the poor beasts die by hundreds. One can hardly conceive how it happens, notwithstanding these many deaths and the immense numbers that are murdered almost daily on these boundless wastes called prairies, besides the hosts that are drowned in the freshets, and the hundreds of young calves who die in early spring, so many are yet to be found. Daily we see so many that we hardly notice them more than the cattle in our pastures about our homes. But this cannot last; even now there is a perceptible difference in the size of the herds, and before many years the Buffalo, like the Great Auk, will have disappeared; surely this should not be permitted.

*Source*: John James Audubon, "The Missouri River Journals," in Maria R. Audubon, *Audubon and His Journals*, Vol. II (New York: Scribner's, 1899), pp. 107, 131.

---

## DOCUMENT 36: Ralph Waldo Emerson on Nature (1844, 1884)

The leader of a group of New England idealists known as the transcendentalists, Ralph Waldo Emerson called on people to give heed to the relationship between humanity and nature. The transcendentalists, who were influenced by German idealist philosophers and American romantic writers and artists, were heirs to a uniquely American Protestant view of nature that was firmly rooted in the writings of eighteenth-century American theologians such as Jonathan Edwards [see Document 16].

### A. From *Essay on Nature*, 1844

To speak truly, few adult persons can see nature. Most persons do not see the sun. At least they have a very superficial seeing. The sun illuminates only the eye of the man, but shines into the eye and the heart of the child. The lover of nature is he whose inward and outward senses are still truly adjusted to each other; who has retained the spirit of infancy even into the era of manhood. His intercourse with heaven and earth becomes part of his daily food. In the presence of nature a wild delight runs through the man, in spite of real sorrows. Nature says,—he is my creature and maugre all his impertinent griefs, he shall be glad with me. Not the sun or the summer alone, but every hour and season yields its tribute of delight; for every hour and change corresponds to and authorizes a different state of the mind, from breathless noon to grimmest midnight. Nature is a setting that fits equally well a comic or a mourning piece. In good health, the air is a cordial of incredible virtue. Crossing a bare common, in snow puddles, at twilight, under a clouded sky, without having in my thoughts any occurrence of special good fortune, I have enjoyed a perfect exhilaration. I am glad to the brink of fear. In the woods, too, a man casts off his years, as the snake his slough, and at what period soever of life, is always a child. In the woods is perpetual youth. Within these plantations of God, a decorum and sanctity reign, a perennial festival is dressed, and the guest sees not how he should tire of them in a thousand years. In the woods, we return to reason and faith. There I feel that nothing can befall me in life,—no disgrace, no calamity (leaving me my eyes), which nature cannot repair. Standing on the bare ground,—my head bathed by the blithe air, and uplifted into infinite space,—all mean egotism vanishes. I become a transparent eye-ball; I am nothing; I see all; the currents of the Universal Being circulate through me; I am part or parcel of God. The name of the nearest friend sounds then foreign and accidental: to be brothers, to be acquaintance,—master or servant, is then a trifle and a disturbance. I am the lover of uncontained and immortal beauty. In the wilderness, I find something more dear and connate than in streets or villages. In the tranquil landscape, and especially in the distant line of the horizon, man beholds somewhat as beautiful as his own nature.

The greatest delight which the fields and woods minister is the suggestion of an occult relation between man and the vegetable. I am not alone and unacknowledged. They nod to me, and I to them. The waving of the boughs in the storm is new to me and old. It takes me by surprise, and yet is not unknown. Its effect is like that of a higher thought or a better emotion coming over me, when I deemed I was thinking justly or doing right.

Yet it is certain that the power to produce this delight does not reside in nature, but in man, or in a harmony of both.

### B. From *The American Scholar*, 1884

The first in time and the first in importance of the influences upon the mind is that of nature. Every day, the sun; and, after sunset, Night and her stars. Ever the winds blow; ever the grass grows. Every day, men and women, conversing, beholding and beholden. The scholar is he of all men whom this spectacle most engages. He must settle its value in his mind. What is nature to him? There is never a beginning, there is never an end, to the inexplicable continuity of this web of God, but always circular power returning into itself. Therein it resembles his own spirit, whose beginning, whose ending, he never can find,—so entire, so boundless. Far too as her splendors whine, system on system shooting like rays, upward, downward, without centre, without circumference,—in the mass and in the particle, Nature hastens to render account of herself to the mind.

*Source*: Ralph Waldo Emerson, *Nature: Addresses and Lectures* (Boston: Houghton Mifflin, 1884), pp. 14–17, 86–87.

---

## DOCUMENT 37: William Cullen Bryant's Proposal for a Great Municipal Park (1844)

For nearly three-quarters of a century, beginning in 1811 with his poem "Thanatopsis," William Cullen Bryant inspired American nature lovers, artists, and writers by conveying a sense of the wonder and divinity of nature. In 1844, inspired by the great parks of Europe, including Regents Park in London, the noted poet and editor of the influential *New York Evening Post* proposed setting aside a very large tract of land for a municipal park. Previously, half a dozen or so acres had been set aside for local parks, such as Madison Square in New York City, but no municipal public park on the scale proposed existed anywhere in the world, for the great European parks were actually private lands that had been opened to the public. Although the site he suggested was not the one finally selected for New York's Central Park, he set in motion a movement to create a rural park in the city.

If the public authorities, who expend so much of our money in laying out the city, would do what is in their power, they might give our vast population an extensive pleasure ground for shade and recreation in these sultry afternoons, which we might reach without going out of town. . . .

On the road to Harlem, between Sixty-eighth Street on the south, and Seventy-seventh on the north, and extending from Third Avenue to the East River, is a tract of beautiful woodland, comprising sixty or seventy acres, thickly covered with old trees, intermingled with a variety of shrubs. The surface is varied in a very striking and picturesque manner, with craggy eminences, and hollows, and a little stream runs through the midst. The swift tides of the East River sweep its rocky shores, and the fresh breeze of the bay comes in, on every warm summer afternoon, over the restless waters. The trees are of almost every species that grows in our woods—the different varieties of ash, the birch, the beech, the linden, the mulberry, the tulip tree, and others; the azalea, the kalmia, and other flowering shrubs are in bloom here in their season, and the ground in spring is gay with flowers. There never was a finer situation for the public garden of a great city. Nothing is wanting but to cut winding paths through it, leaving the woods as they now are, and introducing here and there a jet from the Croton aqueduct, the streams from which would make their own waterfalls over the rocks, and keep the brooks running through the place always fresh and full.

*Source*: William Cullen Bryant, "A New Park," *New York Evening Post*, July 3, 1844, quoted in Allan Nevins, *The Evening Post: A Century of Journalism* (New York: Russell & Russell, 1968; reissue of 1922 edition), p. 194.

---

## DOCUMENT 38: Andrew Jackson Downing Talks about Public Parks and Gardens (1848)

The horticulturalist, nurseryman, and landscape architect Andrew Jackson Downing is frequently given credit for instigating the urban parks movement, but that honor actually belongs to William Cullen Bryant [see Document 37]. Downing, however, was the first to view the building of a great urban park as consonant with American democracy. In this selection Downing discusses the American rural cemeteries that were the precursors of the great municipal parks: Mt. Auburn, opened in Cambridge, Massachusetts, in 1832; Laurel Hill, established on the outskirts of Philadelphia in 1836; and Green-wood, opened in Brooklyn, New York, in 1840. The popularity of these cemeteries—by 1852 100,00 people were visiting Green-wood annually—gave impetus to the urban parks movement.

*Traveller*. I dare say you will be surprised to hear me say that the French and Germans—difficult as they find it to be republican, in a political sense—are practically far more so, in many of the customs of *social* life, than Americans.

*Editor*. Such as what, pray?

*Trav*. Public enjoyments, open to all classes of people, provided at public cost, maintained at public expense, and enjoyed daily and hourly by all classes of persons.

*Ed*. Picture galleries, libraries, and the like, I suppose you allude to?

*Trav*. Yes; but more especially at the present moment, I am thinking of PUBLIC PARKS and GARDENS—those salubrious and wholesome breathing places, provided in the midst of, or upon the suburbs of so many towns on the continent—full of really grand and beautiful trees, fresh grass, fountains, and, in many cases, rare plants, shrubs and flowers. Public picture galleries, and even libraries, are intellectual luxuries; and though we must and will have them, as wealth accumulates, yet I look upon public parks and gardens, which are great social enjoyments, as naturally coming first. Man's social nature stands before his intellectual one in the *order* of cultivation.

*Ed.* But these great public parks are mostly the appendages of royalty, and have been created for purposes of show and magnificence, quite incompatible with our ideas of republican simplicity.

*Trav.* Not at all. In many places these parks were made for royal enjoyment; but even in these, they are, on the continent, no longer held for royal use, but are the pleasure grounds of the public generally. Look, for example, at the Garden of the Tuileries—spacious, full of flowers, green lawns, orange trees and rare plants, in the very heart of Paris, and all open to the public, without charge.

\* \* \*

*Ed.* Enough. I am fully satisfied of the benefits of these places of healthful public enjoyment, and of their being most completely adapted to our institutions. But how to achieve them? What do we find among us to warrant a belief that public parks, for instance, are within the means of our people?

*Trav.* Several things: but most of all, the condition of our public *cemeteries* at the present moment. Why, twenty years ago, such a thing as an embellished, rural cemetery was unheard of in the United States; and at the present moment, we surpass all other nations in these beautiful resting places for the dead. Green-wood, Mount Auburn, and Laurel Hill, are as much superior to the far famed *Père la Chaise* of Paris, in natural beauty, tasteful arrangement, and all that constitutes the charm of such a spot, as St. Peter's is to the Boston State House. Indeed, these cemeteries are the only places in the country that can give an untravelled American any idea of the beauty of many of the public parks and gardens abroad. Judging from the crowds of people in carriages, and on foot, which I find constantly thronging Green-wood and Mount Auburn, I think it is plain enough how much our citizens, of all classes, would enjoy public parks on a similar scale.

*Source*: Andrew Jackson Downing, "A Talk about Public Parks and Gardens," *Horticulturalist* 3, no. 4 (October 1848): 154, 157.

## DOCUMENT 39: Swamp and Overflow Act (1850)

The passage of the Swamp and Overflow Act reflected both a lack of understanding of the nature and significance of wetlands by Americans and the ever-present desire for more land for farmers and real estate developers. This act made the development of the Florida Everglades possible and also led to some classic land frauds in the 1930s and 1940s. Its passage was followed by a century of profligate wetland draining that dried up more than half of the nation's invaluable flood moderators and biological gold mines. Much of the "reclamation" attitude still persists, and conflict over the protection and restoration of wetland areas like the Everglades continues to this day.

*Be it enacted by the Senate and House of Representatives of the United States of America in Congress assembled,* That to enable the State of Arkansas to construct the necessary levees and drains to reclaim the swamp and over-flowed lands therein, the whole of those swamp and overflowed lands, made unfit thereby for cultivation, which shall remain unsold at the passage of this act, shall be, and the same are hereby granted to said State.

*And be it further enacted,* That it shall be the duty of the Secretary of the Interior, as soon as may be practicable after the passage of this act, to make out an accurate list and plats of the lands described as aforesaid and transmit the same to the governor of the State of Arkansas, and, at the request of said governor, cause a patent to be issued to the State therefor; and on that patent, the fee simple to said lands shall vest in the State of Arkansas, subject to the disposal of the legislature thereof: *Provided, however,* That the proceeds of said lands, whether from sale or by direct appropriation in kind, shall be applied, exclusively, as far as necessary, to the purpose of reclaiming said lands by means of the levees and drains aforesaid.

*And be it further enacted,* That in making out a list and plat of the land aforesaid, the greater part of which is "wet and unfit for cultivation," shall be included in said list and plats; but when the greater part of a subdivision is not of that character, the whole of it shall be excluded therefrom.

*And be it further enacted,* That the provisions of this act be extended to, and their benefits be conferred upon, each of the other States of the Union in which such swamp and overflowed lands, known as designated as aforesaid, may be situated.

*Source:* George Minot, ed., *The Statutes at Large and Treaties of the United States of America,* Vol. 9 (Boston: Little, Brown, 1854), 31st Cong., 1st sess., chap. 84, September 28, 1850, p. 519.

## DOCUMENT 40: The Shattuck Report's Recommendations for Sanitary Improvement (1850)

The degradation of sanitary conditions and the high death rate among the poor that accompanied the rapid growth of urban populations spurred various state and local organizations to examine the spread of communicable diseases as a result of unsanitary water, milk, food, waste disposal, and living conditions and to make recommendations for change. The Massachusetts Sanitary Commission's Report, prepared under the direction of Lemuel Shattuck, a pioneer in American public health, was one of the most thorough and influential of these studies. The conditions it sought to address were well documented in both the fiction [see Document 41] and the nonfiction [see Document 46] of the period. While many state and local authorities had begun to deal with these problems by the mid-eighteenth century, several of the problems proved overwhelming for local authorities, and as the century progressed, they intensified [see Documents 64, 65, 67, and 72]. The federal government did not start to confront many of the issues raised in the Shattuck Report until well into the twentieth century.

We recommend that a GENERAL BOARD OF HEALTH be established, which shall be charged with the general execution of the laws of the State, relating to the enumeration, the vital statistics, and the public health of the inhabitants. . . .

We recommend that a LOCAL BOARD OF HEALTH be appointed in every city and town. . . .

We recommend that the laws relating to the public registration of births, marriages, and deaths be perfected and carried into effect in every city and town of the State. . . .

We recommend that, in laying out new towns and villages, and in extending those already laid out, ample provision be made for a supply, in purity and abundance, of light, air, and water; for drainage and sewerage; for paving and for cleanliness. . . .

We recommend that, before erecting any new dwelling-house, manufactory, or other building, for personal accommodation, either as a lodging-house or place of business, the owner or builder be required to give notice to the local Board of Health, of his intention and of the sanitary arrangements he proposes to adopt. . . .

We recommend that local Boards of Health endeavor to prevent or mitigate the sanitary evils arising from overcrowded lodging-houses and cellar-dwellings. . . .

We recommend that open spaces be reserved, in cities and villages,

for public walks; that wide streets be laid out; and that both be orna-
mented with trees. . . .

We recommend that local Boards of Health, and other persons interested, endeavor to ascertain, by exact observation, the effect of mill-
ponds, and other collections or streams of water, and of their rise and
fall, upon the health of the neighboring inhabitants. . . .

We recommend that the local Boards of Health provide for periodical
house-to-house visitation, for the prevention of epidemic diseases, and
for other sanitary purposes. . . .

We recommend that measures be taken to ascertain the amount of
sickness suffered in different localities; and among persons of different
classes, professions, and occupations. . . .

We recommend that nuisances endangering human life or health, be
prevented, destroyed, or mitigated. . . .

We recommend that measures be taken to prevent or mitigate the sanitary evils arising from the use of intoxicating drinks, and from haunts
of dissipation. . . .

We recommend that the authority to make regulations for the quar-
antine of vessels be intrusted to the local Boards of Health. . . .

We recommend that measures be adopted for preventing or mitigating
the sanitary evils arising from foreign emigration. . . .

We recommend that public bathing-houses and wash-houses be established in all cities and villages. . . .

We recommend that, whenever practicable, the refuse and sewage of
cities and towns be collected, and applied to the purposes of agriculture. . . .

We recommend that measures be taken to prevent, as far as practicable, the smoke nuisance. . . .

We recommend that the sanitary effects of patent medicines and other
nostrums, and secret remedies, be observed; that physicians in their pre-
scriptions and names of medicines, and apothecaries in their compounds,
use great caution and care; and that medical compounds advertised for
sale be avoided, unless the material of which they are composed be
known, or unless manufactured and sold by a person of known honesty
and integrity. . . .

We recommend that local Boards of Health, and others interested, endeavor to prevent the sale and use of unwholesome, spurious, and adulterated articles, dangerous to the public health designed for food, drink,
or medicine.

*Source*: Lemuel Shattuck et al., *Report of the Sanitary Commission of Massachusetts*
(Cambridge, MA: Harvard University Press, 1948; facsimile of *Report of a General
Plan for the Promotion of Public and Personal Health* [Boston: Dutton & Wentworth,
1850]), pp. 111, 115, 135, 153, 164, 166, 168, 171, 183, 200, 209, 212, 218, 220.

## DOCUMENT 41: Rebecca Harding Davis on Smoke and Soot in a Mill Town (1861)

This fictional account of life in a mill town is an indictment of air pollution caused by the burning of coal as well as the general squalor of the lives of poor immigrants.

A cloudy day: do you know what that is in a town of iron-works? The sky sank down before dawn, muddy, flat, immovable. The air is thick, clammy with the breath of crowded human beings. It stifles me. I open the window, and looking out, can scarcely see through the rain the grocer's shop opposite, where a crowd of drunken Irishmen are puffing Lynchburg tobacco in their pipes. I can detect the scent through all the foul smells ranging loose in the air.

The idiosyncrasy of this town is smoke. It rolls sullenly in slow folds from the great chimneys of the iron-foundries, and settles down in black, slimy pools on the muddy streets. Smoke on the wharves, smoke on the dingy boats, on the yellow river,—clinging in a coating of greasy soot to the house-front, the two faded poplars, the faces of the passers-by. The long train of mules, dragging masses of pig-iron through the narrow street, have a foul vapor hanging to their reeking sides. Here, inside, is a little broken figure of an angel pointing upward from the mantel-shelf; but even its wings are covered with smoke, clotted and black. Smoke everywhere! A dirty canary chirps desolately in a cage beside me. Its dream of green fields and sunshine a very old dream,—almost worn out, I think.

*Source*: Rebecca Harding Davis, "Life in the Iron-Mills," *Atlantic Monthly*, 7 (April 1861): 430.

## DOCUMENT 42: Homestead Act (1862)

The object of the Homestead Act was to encourage the rapid settlement of the western territories that the United States had acquired. Although some of these lands were unsuitable for intensive farming, farmers who had exhausted their land in the more settled areas of the United States were lured west by the availability of free land. The intensive cultivation of semiarid lands would eventually create a host of problems for the farmers and for the nation as a whole.

*Be it enacted . . .* , That any person who is the head of a family, or who has arrived at the age of twenty-one years, and is a citizen of the United States, or who shall have filed his declaration of intention to become such, as required by the naturalization laws of the United States, and who has never borne arms against the United States Government or given aid and comfort to its enemies, shall, from and after the first January, eighteen hundred and sixty-three, be entitled to enter one quarter section or a less quantity of unappropriated public land, upon which said person may have filed a preemption claim, or which may, at the time the application is made, be subject to preemption at one dollar and twenty-five cents, or less, per acre; . . . *Provided*, That any person owning or residing on land may, under the provisions of this act, enter other land lying contiguous to his or her said land, which shall not, with the land so already owned and occupied, exceed in the aggregate one hundred sixty acres.

*And be it further enacted*, That the person applying for the benefit of this act shall, upon application to the register of the land office in which he or she is about to make such entry, make affidavit before the said register or receiver that he or she is the head of a family, or is twenty-one years or more of age, or shall have performed service in the army or navy of the United States, and that he has never borne arms against the Government of the United States or given aid and comfort to its enemies, and that said application is made for his or her exclusive use and benefit, and that said entry is made for the purpose of actual settlement and cultivation, and not either directly or indirectly for the use or benefit of any other person or persons whomsoever; and upon filing the said affidavit with the register or receiver, and on payment of ten dollars, he or she shall thereupon be permitted to enter the quantity of land specified.

*Source*: George P. Sanger, ed., *The Statutes at Large, Treaties, and Proclamations of the United States of America*, Vol. 12, Part 2 (Boston: Little Brown, 1865), 37th Cong., 2nd sess., chap. 75, March 20, 1862, p. 392.

---

## DOCUMENT 43: George Perkins Marsh's *Man and Nature* (1864)

George Perkins Marsh, a lawyer and philologist who served for a term as a U.S. congressman, was the first to put forth the concept of the "carrying capacity" of the land and to point out that human activity could cause permanent change in the land. His book *Man and Nature*, from which this selection is taken, offers a broad view of the impact of

human activity on nature and the balance of life. It had a major influence on most of the naturalists (who would now be called ecologists)
of the end of the nineteenth and the beginning of the twentieth centuries. Marsh's wide-angle perspective resulted not only from his varied
experiences in the United States but also from his many years in Europe, during which he served as ambassador to Turkey and to Italy.

In the rudest stages of life, man depends upon spontaneous animal
and vegetable growth for food and clothing, and his consumption of
such products consequently diminishes the numerical abundance of the
species which serve his uses. At more advanced periods, he protects and
propagates certain esculent vegetables and certain fowls and quadrupeds, and, at the same time, wars upon rival organisms which prey upon
these objects of his care or obstruct the increase of their numbers. Hence
the action of man upon the organic world tends to subvert the original
balance of its species, and while it reduces the numbers of some of them,
or even extirpates them altogether, it multiplies other forms of animal
and vegetable life.

The extension of agricultural and pastoral industry involves an enlargement of the sphere of man's domain, by encroachment upon the
forests which once covered the greater part of the earth's surface otherwise adapted to his occupation. The felling of the woods has been attended with momentous consequences to the drainage of the soil, to the
external configuration of its surface, and probably, also, to local climate;
and the importance of human life as a transforming power is, perhaps,
more clearly demonstrable in the influence man has thus exerted upon
superficial geography than in any other result of his material effort.

Lands won from the woods must be both drained and irrigated; river
banks and maritime coasts must be secured by means of artificial bul-
warks against inundation by inland and by ocean floods; and the needs
of commerce require the improvement of natural, and the construction
of artificial channels of navigation. Thus man is compelled to extend over
the unstable waters the empire he had already founded upon the solid
land.

The upheaval of the bed of seas and the movements of water and of
wind expose vast deposits of sand, which occupy space required for the
convenience of man, and often, by the drifting of their particles, over-
whelm the fields of human industry with invasions as disastrous as the
incursions of the ocean. On the other hand, on many coasts, sand hills
both protect the shores from erosion by the waves and currents, and
shelter valuable grounds from blasting sea winds. Man, therefore, must
sometimes resist, sometimes, promote, the formation and growth of
dunes, and subject the barren and flying sands to the same obedience to
his will to which he has reduced other forms of terrestrial surface.

Besides these old and comparatively familiar methods of material improvement, modern ambition aspires to yet grander achievements in the conquest of physical nature, and projects are meditated which quite eclipse the boldest enterprises hitherto undertaken for the modification of geographical surface.

*Source*: George P. Marsh, *Man and Nature; or, Physical Geography as Modified by Human Action* (New York: Scribner, 1864), pp. iii–v.

## DOCUMENT 44: Henry David Thoreau on the Value of Living Things (1864)

Henry David Thoreau (1817–1862), a disciple of Ralph Waldo Emerson [see Document 36] and associated with the transcendentalists, was very much an original thinker. His writings about the importance of leaving nature undisturbed, the need for all humans to have contact with nature, and the relationship between humans and other living things were not fully appreciated until the mid-twentieth century, when environmentalists canonized him as their patron saint. He is best known for his essay *Walden*, a record of the two years he spent living in a cabin near Walden Pond in Concord, Massachusetts. In the following passage from a journal he kept during a trip to Maine, Thoreau expresses dismay over the human use of other living things. He also offers a prescient model for a U.S. national parks system.

Strange that so few ever come to the woods to see how the pine lives and grows and spires, lifting its evergreen arms to the light,—to see its perfect success; but most are content to behold it in the shape of many broad boards brought to market, and deem *that* its true success! But the pine is no more lumber than man is, and to be made into boards and houses is no more its true and highest use than the truest use of a man is to be cut down and made into manure. There is a higher law affecting our relation to pines as well as to men. A pine cut down, a dead pine, is no more a pine than a dead human carcass is a man. Can he who has discovered only some of the values of whalebone and whale oil be said to have discovered the true use of the whale? Can he who slays the elephant for his ivory be said to have "seen the elephant"? These are petty and accidental uses; just as if a stronger race were to kill us in order to make buttons and flageolets of our bones; for everything may serve a lower as well as a higher use. Every creature is better alive than dead, men and moose and pine-trees, and he who understands it aright will rather preserve its life than destroy it.

* * *

The kings of England formerly had their forests "to hold the king's game," for sport or food, sometimes destroying villages to create or extend them; and I think that they were impelled by a true instinct. Why should not we, who have renounced the king's authority, have our national preserves, where no villages need be destroyed, in which the bear and panther, and some even of the hunter race, may still exist, and not be "civilized off the face of the earth,"—our forests, not to hold the king's game merely, but to hold and preserve the king himself also, the lord of creation,—not for idle sport or food, but for inspiration and our own true recreation? Or shall we, like the villains, grub them all up, poaching on our own national domains?

*Source*: Henry David Thoreau, *The Maine Woods* (Boston: Houghton, Mifflin, 1893), pp. 163–64, 212–13.

---

## DOCUMENT 45: Act Granting Yo-Semite Valley to California (1864)

As the first large public park in the United States, Yosemite became a model for future parks in the national parks system, which was established a few years later. The federal stipulations regarding the use of the land emphasized the public nature of the park and made clear the intent to establish a permanent park.

*Be it enacted* . . . , That there shall be, and is hereby, granted to the State of California the "Cleft" or "Gorge" in the granite peak of the Sierra Nevada mountains, situated in the county of Mariposa in the State aforesaid, and the headwaters of the Merced River, and known as the Yo-Semite valley, with its branches or spurs, in estimated length fifteen miles, and in average width one mile back from the main edge of the precipice, on each side of the valley, with the stipulation, nevertheless, that the said State shall accept this grant upon the express conditions that the premises shall be held for public use, resort, and recreation; shall be inalienable for all time; but leases not exceeding ten years may be granted for portions of said premises. All incomes derived from leases of privileges to be expended in the preservation and improvement of the property, or the roads leading thereto; the boundaries to be established at the cost of said State by the United States surveyor-general of California, whose official plat, when affirmed by the commissioner of the general land-office, shall constitute the evidence of the locus, extent, and limits of the said Cleft or Gorge; the premises to be managed by the governor of the State with eight other commissioners, to be appointed by the executive of California, and who shall receive no compensation for their services.

*Source: United States Statutes at Large*, Vol. 13 (Boston: Little, Brown, 1866), 38th Cong., 1st sess., chap. 184, June 30, 1864, p. 325.

## DOCUMENT 46: The Citizens' Association of New York on Sewage and Disease (1865)

In the latter part of the nineteenth century numerous civic associations were formed to advocate for improvements in the local environment. Their newsletters railed against poor sanitation as well as noise and air pollution. In time, some of these organizations became strong supporters of the urban parks movement.

The unspeakable filthiness and neglect of the privies pertaining to the tenant-houses demand attention. These necessaries of every domicile are so neglected and filthy in all the crowded districts of the city as to have become prolific sources of obstinate and fatal maladies of a diarrhoeal and febrile character, and they must be reckoned among the most active of localizing causes of prevailing diseases among the poor. The miserable economy that has attached to every tenant-house, court, or cellar a series of *midden* sinks, frequently without any sewer connection, and seldom with sufficient drainage of any kind, should be superseded by suitable water-closet arrangements for constant "flushing" and cleanliness. Reform in these matters is vitally important to the health of tenant-houses.

*Source: Report of the Council of Hygiene and Public Health of the Citizens' Association of New York upon the Sanitary Condition of the City* (New York: Appleton, 1865), p. xci.

## DOCUMENT 47: John Muir on the Spirituality of Nature (1866)

John Muir, an amateur naturalist noted for his books and magazine articles on the mountains, valleys, and parks of the West, was a supporter of wilderness preservation in the western states [see Document 68]. He first achieved national attention as a result of a letter about the calypso borealis, a rare white orchid, that was quoted in the *Boston Recorder*. The letter recounted the turning point in Muir's life that moved him to become what today would be termed an "advocate of the rights of nature" or biocentrist. His description of the spirituality of the encounter with natural beauty echoes the experiences of Jonathan Edwards [see Document 16] and the New England transcendentalists.

For several days in June I had been forcing my way through woods that seemed to become more and more dense, and among bogs more and more difficult to cross, when, one warm afternoon, after descending a hillside covered with huge half-dead hemlocks, I crossed an ice-cold stream, and espied two specimens of Calypso. There upon an open plat of yellow moss, near an immense rotten log, were these little plants so pure.

They were alone. Not a vine was near, nor a blade of grass, nor a bush. Nor were there any birds or insects, for great blocks of ice lay screened from the summer's sun by deep beds of moss, and chilled the water. They were indeed alone, for the dull ignoble hemlocks were not companions, nor was the nearer arbor-vitae, with its root-like pendulous branches decaying confusedly on the wet, cold ground.

I never before saw a plant so full of life; so perfectly spiritual, it seemed pure enough for the throne of its Creator. I felt as if I were in the presence of superior beings who loved me and beckoned me to come. I sat down beside them and wept for joy. Could angels in their better land show us a more beautiful plant? How good is our Heavenly Father in granting us such friends as are these plant-creatures, filling us wherever we go with pleasure so deep, so pure, so endless.

I cannot understand the nature of the curse, "Thorns and thistles shall it bring forth to thee." Is our world worse for this "thistly curse"? Are not all plants beautiful? or in some way useful? Is our world better for this "thistly curse"? Would not the world suffer by the banishment of a single weed? The curse must be in ourselves.

*Source*: John Muir to Mrs. Jeanne Carr, quoted in J. D. Butler, "The Calypso Borealis: Botanical Enthusiasm," *Boston Recorder*, December 21, 1866, p. 1, in Muir, Scrapbook I, p. 26 (John Muir Collection at the University of the Pacific, Stockton, CA).

---

## DOCUMENT 48: Frederick Law Olmsted and Calvert Vaux on Creating Parks to Serve the Public (1866, 1872)

Frederick Law Olmsted and Calvert Vaux developed their two great New York City parks, Central Park in Manhattan and Prospect Park in

Brooklyn, to fulfill a democratic vision of park space accessible to all classes of people, based in part on a belief that contact with nature has an uplifting effect on the human spirit. Both Olmsted and Vaux had visited Green-wood Cemetery in Brooklyn in 1852 and were very much aware of its popularity among a broad cross-section of the city's population. In the first selection, Olmsted and Vaux justify their approach to park design on the basis of their knowledge of European parks. In the second selection, they detail how they went about translating the European idea of a park into a park suitable for a great city in a democratic nation.

## A. From a Report to the Commissioners of Prospect Park, 1866

The word park has different significations, but that in which we are now interested has grown out of its application centuries ago, simply to hunting grounds; the choicest lands for hunting grounds being those in which the beasts of the chase were most happy, and consequently most abundant, sites were chosen for them, in which it was easy for animals to turn from rich herbage to clear water, from warm sunlight to cool shade; that is to say, by preference, ranges of well-watered dale-land, broken by open groves and dotted with spreading trees, undulating in surface, but not rugged. Gay parties of pleasure occasionally met in these parks, and when these meetings occurred the enjoyment otherwise obtained in them was found to be increased. Hence, instead of mere hunting lodges and hovels for game-keepers, extensive buildings and other accommodations, having frequently a festive character, were after a time provided within their enclosures. Then it was found that people took pleasure in them without regard to the attractions of the chase, or of conversation and this pleasure was perceived to be, in some degree, related to their scenery, and in some degree to the peculiar manner of association which occurred in them; and this was also found to be independent of intellectual gifts, tranquilizing and restorative to the powers most tasked in ordinary social duties, and stimulating only in a healthy and recreative way to the imagination. Hence, after a time, parks began to be regarded and to be maintained with reference, more than any thing else, to the convenient accommodation of numbers of people, desirous of moving for recreation among scenes that should be gratifying to their taste or imagination.

In the present century, not only have the old parks been thus maintained, but many new parks have been formed with these purposes exclusively in view, especially within and adjoining considerable towns and it is upon our knowledge of these latter that our simplest conception of a town park is founded. It is from experience in these that all our ideas of parks must spring.

## B. From a Letter to H. G. Stebbins (president of the New York City Department of Parks), 1872

[One of the changes in the plan of Central Park] was suggested by the frightful increase of mortality among very young children which annually occurs in this city about mid-summer; the number of deaths of infants, notwithstanding so many are taken out of town, often being double as many in a day about the middle of July as in any day of several previous months. The causes act in part directly upon the children, but largely, also indirectly, by inducing nervous irritation with nursing mothers.

A visit to the country offers the surest means of escaping the danger, and, in incipient stages, the best means of cure of the special disorders in which the danger lies. To most mothers, however, this is impracticable, and the best that can be done is to spend an occasional day or part of a day on the Park. It has been for some years a growing practice with physicians to advise this course.

The whole Park is, of course, open as much to mothers with children as to any other class; but on a hot day a mother carrying a sick child, and perhaps leading other children, if she follows the throng, is liable to become more heated and feverish through fatigue, anxiety and various slight embarrassments, than if she remained quietly within a close, dark chamber. If she comes with a party of friends, she will be glad to find some quiet nook in which, while others wander, she can be left with her baby. The class of considerations thus suggested had influenced the treatment of several localities, but had been controlling in a larger way than elsewhere at the point in question.

. . . [J]ust here in the midst of the general bleakness, barrenness and filth of this quarter of the Park site, there was a pretty bit of natural scenery, having a somewhat wild and secluded character. It was designed to follow up the natural suggestions of this class, and by thickening and extending the original sylvan defences, secure a more decided effect of rural retirement.

The [natural] advantage for this purpose supplied one ground for the selection of the spot, the proximity of the play-grounds for larger children, another; and that of one of the sunken roads of the Park another; but the main reason for it was the fact that *it was the precise point in the Park which could be reached with the fewest steps* on an average, by visitors coming from the denser parts of the city by seven different lines of railway, and after the Park should be entered, wholly along walks by which the crossing of any carriage road would be avoided.

*Source*: **A**. Report to Commissioners of Prospect Park, 1866, and **B**. Department of Public Parks, 2nd Annual Report, Appendix B, in Frederick Law Olmsted, Jr., and Theodora Kimball, eds., *Frederick Law Olmsted: Landscape Architect, 1822–1903*

(New York: Benjamin Blom, 1970; reissue of Olmsted and Kimball, *Forty Years of Landscape Architecture: Being the Professional Papers of Frederick Law Olmsted, Senior* [1928]), pp. 211–12, 242–43.

---

## DOCUMENT 49: Charles Darwin on the Similarity between Humans and Other Animals (1871)

Charles Darwin, like many other early British naturalists, was educated for the ministry but had been fascinated by nature since childhood. His development of the theory of natural selection was influenced in part by a reading of Malthus's *Essay on the Principle of Population* [see Document 26] and a contemplation of the effects of living in an over-crowded world. His carefully documented theoretical work presented in *The Origin of Species* launched a continuing debate about whether humans are a special creation different from all other living things. The idea that humans are subject to the same "principles of evolution" as other animals was barely alluded to in *The Origin of Species*; it was Thomas Henry Huxley's *Evidence of Man's Place in Nature*, published in 1863, that provided the first detailed discussion of the theory. Darwin did not expand on the topic until 1871, when he wrote *The Descent of Man*.

He who wishes to decide whether man is the modified descendant of some pre-existing form, would probably first inquire whether man varies, however slightly, in bodily structure and in mental faculties; and if so, whether the variations are transmitted to his offspring in accordance with the laws which prevail with the lower animals. Again, are the variations the result, as far as our ignorance permits us to judge, of the same general causes, and are they governed by the same general laws, as in the case of other organisms; for instance, by correlation, the inherited effects of use and disuse, etc.? Is man subject to similar malconformations, the result of arrested development, of reduplication of parts, etc., and does he display in any of his anomalies reversion to some former and ancient type of structure? It might also naturally be inquired whether man, like so many other animals, has given rise to varieties and sub-races, differing but slightly from each other, or to races differing so much that they must be classed as doubtful species? How are such races distributed over the world; and how, when crossed, do they react on each other in the first and succeeding generations? And so with many other points.

The inquirer would next come to the important point, whether man tends to increase at so rapid a rate, as to lead to occasional severe struggles for existence; and consequently to beneficial variations, whether in body or mind, being preserved, and injurious ones eliminated. Do the races or species of men, whichever term may be applied, encroach on and replace one another, so that some finally become extinct? We shall see that all these questions, as indeed is obvious in respect to most of them, must be answered in the affirmative, in the same manner as with the lower animals. . . .

It is notorious that man is constructed on the same general type or model as other mammals. All the bones in his skeleton can be compared with corresponding bones in a monkey, bat, or seal. So it is with his muscles, nerves, blood-vessels, and internal viscera. The brain, the most important of all the organs, follow the same law, as shown by [Thomas Henry] Huxley and other anatomists.

Source: Charles Darwin, *The Descent of Man and Selections in Relation to Sex* (New York: A. L. Burt, n.d.; reprinted from the 2nd English ed. rev.), pp. 5–6.

---

## DOCUMENT 50: Act Establishing Yellowstone National Park (1872)

Scouts and trappers who ventured into the Yellowstone River basin in the early nineteenth century spread stories about beautiful waterfalls, splendid canyons, and spectacular geysers to be found near the headwaters of the Yellowstone and Madison Rivers. In 1870 an expedition led by General Henry Washburne, the surveyor-general of Montana who had served in the Civil War, and Lieutenant Gustavus Doane set out to confirm the truth of these tales. When the Washburne-Doane expedition returned, two members of the group, Cornelius Hedges and Nathaniel Langford, spread the word about the natural wonders of this wild region and generated interest in turning the area into a park that could be enjoyed by the general public.[5] In 1872 President Ulysses S. Grant signed legislation setting aside 2 million acres of federal land primarily in northwestern Wyoming as a park. It was the first time that any central government in the world had designated an area of public land as a permanent park.

*Be it enacted by the Senate and House of Representatives of the United States of America in Congress assembled,* That the tract of land in the Territories of Montana and Wyoming, lying near the head-waters of the Yellowstone river, and described as follows, to wit, commencing at the junction of Gardiner's river with the Yellowstone river, and running east to the meridian passing ten miles to the eastward of the most eastern point of Yellowstone lake; thence south along said meridian to the parallel of latitude passing ten miles south of the most southern point of Yellowstone lake; thence west along said parallel to the meridian passing fifteen miles west of the most western point of Madison lake; thence north along said meridian to the latitude of the junction of the Yellowstone and Gardiner's rivers; thence east to the place of beginning, is hereby reserved and withdrawn from settlement, occupancy, or sale under the laws of the United States, and dedicated and set apart as a public park or pleasuring-ground for the benefit and enjoyment of the people; and all persons who shall locate or settle upon or occupy the same, or any part thereof, except as hereinafter provided, shall be considered trespassers and removed therefrom.

That said public park shall be under the exclusive control of the Secretary of the Interior, whose duty it shall be, as soon as practicable, to make and publish such rules and regulations as he may deem necessary or proper for the care and management of the same. Such regulations shall provide for the preservation, from injury or spoliation, of the timber, mineral deposits, natural curiosities, or wonders within said park and their retention in their natural condition. The secretary may in his discretion, grant leases for building purposes for terms not exceeding ten years, of small parcels of ground, at such places in said park as shall require the erection of buildings for the accommodation of visitors; all of the proceeds of said leases, and all other revenues that may be derived from any source connected with said park, to be expended under his direction in the management of the same, and the construction of roads and bridle-paths therein. He shall provide against the wanton destruction of the fish and game found within said park, and against their capture or destruction for the purposes of merchandise or profit. He shall also cause all persons trespassing upon the same after the passage of this act to be removed therefrom, and generally shall be authorized to take all such measures as shall be necessary or proper to fully carry out the objects and purposes of this act.

*Source: United States Statutes at Large*, Vol. 17 (Boston: Little, Brown, 1873), 42nd Cong., 2nd sess., chap. 24, May 1, 1872, pp. 32–33.

## DOCUMENT 51: Mining Act (1872)

In the 1870s the United States was not only eager to have settlers move into its western territories, it was also anxious to exploit the natural wealth of the region. The giveaway provisions of the Act to Promote the Development of the Mining Resources of the United States, like those of the Homestead Act [see Document 42], were designed to encourage settlement as well as resource development.

The 1872 Mining Act ended the governmental practice, which had been in effect since the colonial period, of charging a royalty for taking minerals from public lands. It allows prospectors and speculators to stake claims to any public lands containing hard rock minerals (i.e., gold and silver as opposed to oil and gas, which are found in shale), even if the land contains only trace amounts of the minerals. As long as the claim holder makes at least $100 worth of improvements on the land annually or obtains a patent for the land, all other individuals are excluded from the land. The Mining Act, with only minor revisions, remains in effect today and continues to govern land use in much of the West. Even its fees remain near the 1872 levels.

*Be it enacted . . .* , That all valuable mineral deposits in lands belonging to the United States, both surveyed and unsurveyed, are hereby declared to be free and open to exploration and purchase, and the lands in which they are found to occupation and purchase, by citizens of the United States and those who have declared their intention to become such, under regulations prescribed by law, and according to the local customs or rules of miners, in the several mining-districts, so far as the same are applicable and not inconsistent with the laws of the United States.

SEC. 2. That mining-claims upon veins or lodes of quartz or other rock in place bearing gold, silver, cinnabar, lead, tin, copper, or other valuable deposits heretofore located, shall be governed as to length along the vein or lode by the customs, regulations, and laws in force at the date of their location. A mining-claim located after the passage of this act, whether located by one or more persons, may equal, but shall not exceed, one thousand five hundred feet in length along the vein or lode; but no location of a mining-claim shall be made until the discovery of the vein or lode within the limits of the claim located. No claim shall extend more than three hundred feet on each side of the middle of the vein at the surface, nor shall any claim be limited by any mining regulation to less than twenty-five feet on each side of the middle of the vein at the surface. . . .

SEC. 3. That the locators of all mining locations heretofore made, or which shall hereafter be made, on any mineral vein, lode, or ledge, situated on the public domain, their heirs and assigns . . . shall have the exclusive right of possession and enjoyment of all the surface included within the lines of their locations, and of all veins, lodes, and ledges throughout their entire depth. . . .

SEC. 5. That the miners of each mining district may make rules and regulations not in conflict with the laws of the United States, or with the laws of the State or Territory in which the district is situated, governing the location, manner of recording, amount of work necessary to hold possession of a mining-claim, subject to the following requirements: The location must be distinctly marked on the ground so that its boundaries can be readily traced. . . . On each claim located after the passage of this act, and until a patent shall have been issued therefor, not less than one hundred dollars' worth of labor shall be performed or improvements made during each year. . . .

SEC. 6. That a patent for any land claimed and located for valuable deposits may be obtained in the following manner: Any person, association, or corporation authorized to locate a claim under this act, having claimed and located a piece of land for such purposes, who has, or have, complied with the terms of this act, may file in the proper land-office an application for a patent . . . and shall thereupon be entitled to a patent for said land in such manner: The register of the land-office, upon the filing of such application, plat, field-notes, notices, and affidavits, shall publish a notice that such application has been made, for the period of sixty days, in a newspaper to be by him designated as published nearest to said claim; and he shall also post such notice in his office for the same period. The claimant at the time of filing this application, or at any time thereafter, within the sixty days of publication, shall file with the register a certificate of the United States surveyor-general that five hundred dollars' worth of labor has been expended or improvements made upon the claim by himself or grantors; that the plat is correct. . . . If no adverse claim shall have been filed with the register and the receiver of the proper land-office at the expiration of the sixty days of publication, it shall be assumed that the applicant is entitled to a patent, upon the payment to the proper officer of five dollars per acre, and that no adverse claim exists.

*Source: United States Statutes at Large*, Vol. 17 (Boston: Little, Brown, 1873), 42nd Cong., 2nd sess., chap. 152, May 10, 1872, p. 91.

---

## DOCUMENT 52: Carl Schurz on the Need for Federal Forest Conservation (1877)

Carl Schurz brought from his native Germany a love of trees and an understanding of forest management. He was the first federal official to recognize the widespread abuse of timbered lands in the United States, and as secretary of the Interior under President Rutherford B. Hayes, he tried (in vain) to control rampant commercial exploitation of federal forests.

The subject of the extensive depredations committed upon the timber on the public lands of the United States has largely engaged the attention of [the Department of the Interior]. That question presents itself in a twofold aspect: as a question of law and as a question of public economy. As to the first point, little need be said. That the law prohibits the taking of timber by unauthorized persons from the public lands of the United States, is a universally known fact. That the laws are made to be executed, ought to be a universally accepted doctrine. That the government is in duty bound to act upon that doctrine, needs no argument. There may be circumstances under which the rigorous execution of a law may be difficult or inconvenient, or obnoxious to public sentiment, or working particular hardship; in such cases it is the business of the legislative power to adapt the law to such circumstances. It is the business of the Executive to enforce the law as it stands.

As to the second point, the statements made by the Commissioner of the General Land Office, in his report, show the quantity of timber taken from the public lands without authority of law to have been of enormous extent. It probably far exceeds in reality any estimates made upon the data before us. It appears, from authentic information before this department, that in many instances the depredations have been carried on in the way of organized and systematic enterprise, not only to furnish timber, lumber, and fire-wood for the home market, but, on a large scale, for commercial exportation to foreign countries.

The rapidity with which this country is being stripped of its forests must alarm every thinking man. It has been estimated by good authority that, if we go on at the present rate, the supply of timber in the United States will, in less than twenty years, fall considerably short of our home necessities. How disastrously the destruction of the forests of a country affects the regularity of the water supply in its rivers necessary for navigation, increases the frequency of freshets and inundations, dries up springs, and transforms fertile agricultural districts into barren wastes is a matter of universal experience the world over. It is the highest time that we should turn our earnest attention to this subject, which so seriously concerns our national prosperity.

The government cannot prevent the cutting of timber on land owned by private citizens. It is only to be hoped that private owners will grow more careful of their timber as it rises in value. But the government can do two things: 1. It can take determined and, as I think, effectual measures to arrest the stealing of timber from public lands on a large scale, which is always attended with the most reckless waste; and, 2. It can preserve the forests still in its possession by keeping them under its control, and by so regulating the cutting and sale of timber on its lands as to secure the renewal of the forest by natural growth and the careful preservation of the young timber.

*Source: Annual Report of the Secretary of the Interior on the Operations of the Depart-ment for the Fiscal Year Ended June 30, 1877* (Washington, D.C.: Government Print-ing Office, 1877), [iii], pp. xv–xx, in Roderick Nash, ed., *Readings in the History of Conservation* (Reading, MA: Addison-Wesley, 1968), pp. 25–26.

---

## DOCUMENT 53: Henry George on Land Development (1879)

---

The social reformer and economist Henry George was working in San Francisco as a printer and editor at the time of the California land boom spurred by the growth of the railroads. He vehemently objected to policies for land development being determined by landowners and speculators.

So far from the recognition of private property in land being necessary to the proper use of land, the contrary is the case. Treating land as pri-vate property stands in the way of its proper use. Were land treated as public property it would be used and improved as soon as there was need for its use or improvement, but being treated as private property, the individual owner is permitted to prevent others from using or im-proving what he cannot or will not use or improve himself. When the title is in dispute, the most valuable land lies unimproved for years; in many parts of England improvement is stopped because, the estates be-ing entailed, no security to improvers can be given; and large tracts of ground which, were they treated as public property, would be covered with buildings and crops, are kept idle to gratify the caprice of the owner. In the thickly settled parts of the United States there is enough land to maintain three or four times our present population, lying un-used, because its owners are holding it for higher prices, and immigrants are forced past this unused land to seek homes where their labor will be far less productive. In every city valuable lots may be seen lying vacant for the same reason. If the best use of land be the test, then private property in land is condemned, as it is condemned by every other con-sideration. It is as wasteful and uncertain a mode of securing the proper use of land as the burning down of houses is of roasting pigs.

*Source*: Henry George, *Progress and Poverty: An Inquiry into the Cause of Industrial Depressions and of Increase of Want with Increase of Wealth* (New York: Robert Schal-kenbach Foundation, 1979), pp. 401–2.

## DOCUMENT 54: Act Establishing the Adirondack Forest Preserve (1885)

"An act to establish a forest commission, and to define its powers and duties and for the preservation of forests" was passed by the New York State legislature on May 15, 1885. In 1894 the "forever wild" clause of the law, section 8, was incorporated into the New York State Constitution as Article VII, section 7: "[T]he lands of the State now owned or hereafter acquired constituting the Forest Preserve as now fixed by law, shall be forever kept as wild forest lands. They shall not be leased, sold or exchanged, or be taken by any corporation, public or private, nor shall the timber thereon be sold, removed or destroyed." More then half a century later, in the 1950s, this clause[6] provided the impetus for Howard Zahnizer, the director of the Wilderness Society, to prepare a legislative draft and for a coalition of environmental groups to run a ten-year campaign [see Document 95] that resulted in passage of the Federal Wilderness Act of 1964. The "forever wild" clause, however, has always had numerous opponents, and in recent years it has been a focal point of efforts to amend the New York State Constitution.

George Bird Grinnell, the editor and owner of *Forest and Stream*, an influential periodical with a broad readership that ranged from outdoorsmen to politicians, was the champion of a variety of conservation causes. He feared that commercial timber interests would be allowed to denude the Adirondacks before the land came under state control.

### A. The Act

SEC. 1. There shall be a forest commission which shall consist of three persons who shall be styled forest commissioners. . . .

SEC. 7. All the lands now owned or which may hereafter be acquired by the state of New York, within the counties of Clinton, excepting the towns of Altona and Dannemora, Essex, Franklin, Fulton, Hamilton, Herkimer, Lewis, Saratoga, St. Lawrence, Warren, Washington, Greene, Ulster and Sullivan, shall constitute and be known as the forest preserve.

SEC. 8. The lands now or hereafter constituting the forest preserve shall be forever kept as wild forest lands. They shall not be sold, nor shall they be leased or taken by any person or corporation, public or private.

SEC. 9. The forest commission shall have the care, custody, control and superintendence of the forest preserve. It shall be the duty of the commission to maintain and protect the forests now on the forest preserve, and to promote as far as practicable the further growth of forests thereon.

It shall also have charge of the public interests of the state, with regard to forests and tree planting, and especially with reference to forest fires in every part of the state. . . . The forest commission may, from time to time, prescribe rules or regulations . . . affecting the whole or any part of the forest preserve, and for its use, care and administration; but neither such rules or regulations, nor anything herein contained shall prevent or operate to prevent the free use of any road, stream or water as the same may have been heretofore used or as may be reasonably required in the prosecution of any lawful business.

### B. George Bird Grinnell's Commentary on the Proposed Act, January 17, 1884

A bill was introduced at Albany last Tuesday by Senator Lansing, which provides for the protection of the Adirondack forests by the establishment of a State Park, to be fenced in and put in charge of a superintendent. The extent of the territory to be included in this park comprises 1,700,000 acres. Of this land the State now owns 750,000, or less than one-half. The bill . . . provides that the State shall assume immediate active control of the forest land now in its possession, and that the remaining 950,000 acres shall come within the same protecting care as it may be gradually abandoned by the present owners and allowed to revert for unpaid taxes.

The bill is a most excellent one, so far as it goes; but it is not sufficient. If the 1,700,000 acres of forest land should be cared for by the State, that care should be assumed at once, before the land has been denuded of its timber. Protection and conservation, now, prompt, adequate—this is what the Adirondack forests demand, not restoration years hence, after the damage shall have been wrought and ruin has followed.

We hear much ado made lest the proposition to assume State control of those lands shall terminate in a huge job; and again we urge that such a fear is not based on good grounds. The forests ought to be saved, even at great (but not exorbitant) cost, and in this day and generation most surely the man with the pocket to fill ought not to stand in the way.

*Source*: **A.** *Laws of New York*, chap. 283, May 15, 1885, pp. 482–83. **B.** "The Adirondacks," *Forest and Stream* 21, no. 25 (January 17, 1884): 489.

---

## DOCUMENT 55: American Ornithologists' Union's Model Law (1886)

---

In 1883 a group of professional and amateur ornithologists chartered the American Ornithologists' Union (AOU) to further the study of bird biology and economics. However, not long after the founding of the AOU, George Bird Grinnell and other members of the group decided to form the Committee on the Protection of North American Birds to effect social action rather than scientific study. The committee's most significant contribution was the writing of the "Model Law," which was widely distributed and used as a prototype for many state bird protection laws.

Section 1.—Any person who shall, within the state of _____, kill any wild bird other than a game-bird, or purchase, offer, or expose for sale any such wild bird, after it has been killed, shall for each offense be subject to a fine of five dollars, or imprisonment for ten days, or both, at the discretion of the court. For the purposes of this act the following only shall be considered game-birds. The Anatidae, commonly known as swans, geese, brant, and river and sea ducks; the Rallidae, commonly known as rails, coots, mud-hens, and gallinules; the Limicolae, commonly known as shore-birds, plovers, surf-birds, snipe, woodcock, sandpipers, tatlers and curlews; the Gallinae, commonly known as wild turkeys, grouse, prairie-chickens, pheasants, partridges, and quails.

Sect. 2.—Any person who shall, within the state of _____, take or needlessly destroy the nest or the eggs of any wild bird, shall be subject for each offense to a fine of five dollars, or imprisonment for ten days, or both, at the discretion of the court.

Sect. 3.—Sections 1 and 2 of this act shall not apply to any person holding a certificate giving the right to take birds, and their nests and eggs, for scientific purposes, as provided for in Section 4 of this act.

*Source*: American Ornithologists' Union, Committee on Protection of Birds, "An Act for the Protection of Birds and Their Nests and Eggs," *Bulletin of the Committee on Protection of Birds*, reprinted in *Science* 7, no. 160 (February 26, 1886): 204.

---

## DOCUMENT 56: George Bird Grinnell and Cecelia Thaxter on the Audubon Society Cause (1886)

Continuing to use *Forest and Stream* to promote his conservation agenda [see Document 54], Grinnell suggested the formation of an association for the protection of birds that would further the work he had started with the AOU [see Document 55]. Grinnell's idea was wildly successful, and within a year he had 39,000 members pledged to protect birds. However, the group, named the Audubon Society, after the ornithologist John Audubon [see Documents 33 and 35], on whose subdivided estate Grinnell had grown up and whose widow had provided his education, was disbanded after two years because Grinnell could not manage the fledgling society along with his weekly journal and publishing business. Nevertheless, within eight years, state Audubon societies had been chartered in Massachusetts and Pennsylvania, and by the turn of the century, many others had been formed. Championing the cause of the Audubon Society were a large number of women, including the writer and gardener Cecelia Thaxter, whose family owned a resort on the Isle of Shoals, off the coast of New Hampshire, that was popular with nature lovers.

## A. George Bird Grinnell's Proposal for Formation of the Audubon Society, February 11, 1886

Very slowly the public are awakening to see that the fashion of wearing the feathers and skins of birds is abominable. There is, we think, no doubt that when the facts about this fashion are known, it will be frowned down and will cease to exist. Legislation of itself can do little against this barbarous practice, but if public sentiment can be aroused against it, it will die a speedy death.

The *Forest and Stream* has been hammering away at this subject for some years, and the result of its blows is seen in the gradual change which has taken place in public sentiment since it began its work. The time has passed for showing that the fashion is an outrageous one, and that it results very disastrously to the largest and most important class of our population—the farmers. These are injured in two ways; by the destruction of the birds, whose food consists chiefly of insects injurious to the growing crops, and of that scarcely less important group the Rapaces, which prey upon the small rodents which devour the crop after it has matured.

The reform in Americas, as elsewhere, must be inaugurated by women, and if the subject is properly called to their notice, their tender hearts will be quick to respond. In England, this matter has been taken up and a widespread interest in it developed. If the women of America will take hold in the same earnest way, they can accomplish an incalculable amount of good.

While individual effort may accomplish much, it will work but slowly, and the spread of the movement will be but gradual. Something more than this is needed. Men, women and children all over our land should take the matter in hand, and urge its importance upon those with whom they are brought in contact. A general effort of this kind will not fail to awaken public interest, and information given to a right-thinking public will set the ball of reform in motion. Our beautiful birds give to many people a great deal of pleasure and add much to the delights of the country. These birds are slaughtered in vast numbers for gain. If the demand for their skins can be caused to fall off, it will no longer repay the bird butchers to ply their trade and the birds will be saved.

. . .

We propose the formation of an association for the protection of wild birds and their eggs, which shall be called the Audubon Society. Its membership is to be free to every one who is willing to lend a helping hand in forwarding the objects for which it is formed. These objects shall be to prevent, so far as possible (1), the killing of any wild birds not used for food; (2) the destruction of nests or eggs of any wild bird, and (3) the wearing of feathers as ornaments or trimming for dress.

### B.   Cecelia Thaxter Attacks Bird-Wearing Women

When the Audubon Society was first organized, it seemed a comparatively simple thing to awaken in the minds of all bird-wearing women a sense of what their "decoration" involved. We flattered ourselves that the tender and compassionate heart of woman would at once respond to the appeal for mercy, but after many months of effort we are obliged to acknowledge ourselves mistaken in our estimate of that universal compassion, that tender heart in which we believed. Not among the ignorant and uncultivated so much as the educated and enlightened do we find the indifference and hardness that baffles and perplexes us. Not always, heaven be praised! But too often,—I think I may say in two-thirds of the cases to which we appeal. One lady said to me, "I think there is a great deal of sentiment wasted on the birds. There are so many of them, they will never be missed any more than mosquitoes. I shall put birds on my new bonnet." . . . and she went her way, a charnel-house of beaks and claws and bones and feathers and glass eyes upon her fatuous head.

Another, mockingly, says, "Why don't you try to save the little fishes in the sea?" and continues to walk the world with dozens of warblers' wings making her headgear hideous. Not one in fifty is found willing to remove at once the birds from her head, even if, languidly, she does acquiesce in the assertion that it is a cruel sin against nature to destroy them. "When these are worn out I am willing to promise not to buy any more," is what we hear, and we are thankful, indeed, for even so much grace; but alas! birds never "wear out."

*Source*: **A.** George Bird Grinnell, "The Audubon Society," *Forest and Stream* 24, no. 3 (February 11, 1886): 41. **B.** Cecelia Thaxter, *Woman's Heartlessness* (Boston 1886; reprinted for the Audubon Society of the State of New York, 1899) in the New York Public Library Audubon Collection.

## DOCUMENT 57: Constitution of the Boone and Crockett Club (1887)

In the latter part of the nineteenth and early part of the twentieth centuries, a number of outdoor clubs devoted to hiking, fishing, and hunting were organized with constitutions that contained conservation agendas aimed at protecting their own interests. Among the most influential of these sportsmen's groups was the Boone and Crockett Club, whose founding members included the painter Albert Bierstadt; George Bird Grinnell; Jay Pierpont; Archibald Rogers, secretary; Theodore Roosevelt, president; and several other Roosevelt family members. William "Buffalo Bill" Cody joined the club soon after its inception.

Article II. The objects of the club shall be—

1. To promote manly sport with the rifle.
2. To promote travel and exploration in the wild and unknown or but partially known portions of the country.
3. To work for the preservation of the large game of this country, and so far as possible, to further legislation for that purpose, and to assist in enforcing the existing laws.
4. To promote inquiry into, and to record observations on the habits and natural history of the various wild animals.
5. To bring about among the members the interchange of opinions and ideas on hunting, travel and exploration of the various kinds of hunting rifles, on the haunts of game animals, etc.

Article III. No one shall be eligible for membership who shall not have killed with the rifle in fair chase, by still-hunting or otherwise, at least one individual of one of the various kinds of American large game.

Article IV. Under the head of American large game are included the following animals: Bear, buffalo (bison), mountain sheep, caribou, cougar, musk ox, white goat, elk (wapiti), wolf (not coyote), pronghorn antelope, moose and deer.

Article V. The term "fair chase" shall not be held to include killing bear, wolf, or cougar in traps, nor "fire-hunting," nor "crusting" moose, elk or deer in deep snow, nor killing game from a boat while it is swimming in the water.

*Source*: "Constitution of the Boone and Crockett Club," in Theodore Roosevelt and George Bird Grinnell, eds., *American Big Game Hunting* (New York: Forest and Stream Publishing, 1893), pp. 337–38.

## DOCUMENT 58: John Wesley Powell on the Lands of the Arid Region (1890)

John Wesley Powell, who was the first person of European descent to travel by boat through the Grand Canyon, spent many years studying the Great Basin region between the Rocky and Sierra mountains while serving on the staff of the U.S. Geological Survey. He was an enthusiastic supporter of the idea of scientifically studying and developing the nation's natural resources.

In 1878, Powell published his original report on the arid regions of West and sent a copy to Carl Schurz. The report was ignored in government circles, however, and Powell spent the next twenty years promoting his ideas in a variety of venues, including popular publications such as *Century Magazine*, from which the following selections are taken. Eventually Powell's writings came to the attention of Theodore Roosevelt, who took many of the ideas for his 1901 address to Congress [see Document 62] from Powell's report. The Reclamation Act [see Document 63], passed the following year, was an outgrowth of the address.

Although Powell's writings on the arid lands show him to be a man of his times who believed that nature could be harnessed and controlled, they also reveal him to have been a visionary conservationist who recognized the need to conserve the forests of the West.

### A. The Irrigable Lands

[W]ill not the hills of New England, the mountains and plains of the sunny South, and the prairies of the middle region be sufficient for the agricultural industries of the United States? The area is vast, the soil is bountiful, and the heavens kindly give their rains. Why should the naked plains and the desert valleys of the far West be redeemed? Why should our civilization enter into a contest with nature to subdue the rivers of the West when the clouds of the East are ready servants?

Gold is found in the graves of the West; silver abounds in the cliffs; copper is found in the mountains; iron, coal, petroleum, and gas are supplied by nature. The mountains and plateaus are covered with stately forest; the climate is salubrious and wonderfully alluring. So the tide of migration rolls westward and the arid region is being carved into States. The people are building cities and towns, erecting factories, and constructing railroads, and great industries of many kinds are already developed. The merchant and his clerk, the banker and his bookkeeper, the superintendent and his operative, the conductor and his brakeman, must be fed; and the men of the West are too enterprising and too industrious to beg bread from the farms of the East. Already they have redeemed more than six million acres of this land; already they are engaged in warfare with the rivers, and have won the first battles. An army of men is enlisted and trained, and they march on a campaign—not for blood, but for bounty; not for plunder, but for prosperity.

But arid lands are not lands of famine, and the sunny sky is not a firmament of devastation. Conquered rivers are better servants than wild clouds. The valleys and plains of the far West have all the elements of fertility that soil can have. As the blood in the body is the stream which supplies the elements of its growth, so the water in the plant is its source of increase. As the body must have more than blood, so the plant must have more than water for its vigorous growth. These conditions of plant growth are light and heat. While the roots of the plant are properly supplied with water and other elements of plant growth, the leaves must be supplied with air and sunshine. The light of a cloudless sky is more invigorating to plants than the gloom of storm. Abundant water and abundant sunshine are the chief conditions for vigorous plant growth, and that agriculture is the most successful which best secures these twin primal conditions; and they are obtained in the highest degree in lands watered by streams and domed by clear skies. For these reasons arid lands are more productive under high cultivation than humid lands. The wheatfields of the desert, the cornfields, the vineyards, the orchards, and the gardens of the far West, far surpass those of the East in luxuriance and productiveness. In the East the field may pine for delayed rains and the green of prosperity fade into sickly saffron, or the vegetation may be beaten down by storms and be drowned by floods; while in the more favored lands of the arid region there is a constant and perfect supply of water by the hand of man, and a constant and perfect supply of sunshine by the economy of nature. The arid lands of the West, last to be redeemed by methods first discovered in civilization, are the best agricultural lands of the continent. Not only must these lands be redeemed because of the wants of the population of that country, they must be redeemed because they are our best lands.

## B. The Nonirrigable Lands

[A]bout one-tenth of the arid region is covered with firewood timber, but this timber is very scant, and often the open spaces are large. It could all stand on one-fiftieth of the entire arid area and not be crowded. The milling timber also covers about another tenth of the ground, but there are many barren places, and usually the trees are widely scattered, so that they could all stand on one-fortieth of the space and still have abundant room. So both classes combined could easily stand on less than one-twentieth of the arid regions.

The merchantable timber is all on the high plateaus and mountains; hence the lands where it grows are not valuable for agricultural purposes. Canyon walls, cliffs, crags, and rocky steeps are not attractive farming-grounds. But more at these great altitudes deep snows fall, ice appears early and lingers long, and frosts come on many a summer night.

The agricultural lands are situate in the valleys where the streams flow. Thus forest and farm are segregated in one region, farming industries in another. It is no small task for the farmer and the villager to haul their wood from distant mountains and to bring poles and logs from the upper region. . . .

The miners are also interested in these forests. As they penetrate with their shafts, drifts, and galleries into the hills and mountains, they carry away to the surface the rock in which the gold, silver, copper, and lead are found, that the metals may be extracted on the ground above.

. . .

Before the white man came the natives systematically burned over the forest lands with each recurrent year as one of their great hunting economies. By this process little destruction of timber was accomplished; but, protected by civilized men, forests are rapidly disappearing. The needles, cones, and brush, together with the leaves of grass and shrubs below, accumulate when not burned annually. New deposits are made from year to year, until the ground is covered with a thick mantle of inflammable material. Then a spark is dropped, a fire is accidentally or purposely kindled, and the flames have abundant food.

There is a practical method by which the forests can be preserved. All of the forest areas that are not dense have some value for pasturage purposes. Grasses grow well in the open grounds, and to some extent among the trees. If herds and flocks crop these grasses, and trample the leaves, and cones into the ground, and make many trails through the woods, they destroy the conditions most favorable to the spread of fire. But if the pasturage is crowded, the young growth is destroyed and the forests are not properly replenished by a new generation of trees. The wooded grounds that are too dense for pasturage should be annually burned over at a time when the inflammable materials are not too dry, so that there may be no danger of great conflagration.

The area of good timber being very small, it has great value, and its rapid destruction is a calamity that cannot well be overestimated. These living forests are always a delight, for in beauty and grandeur they are unexcelled; but dead forests present scenes of desolation that fill the soul with sadness. The vast destruction of values, together with the enormous ravishment of beauty, have for years enlisted the sympathy of intelligent men. Forestry organizations have been formed; conventions have been held; publicists have discussed the subject; and there is a universal sentiment in the West, and a growing opinion in the East, that measures should be taken by the General Government for the protection of the forests. This subject is of profound interest; but sometimes factitious reasons are given which detract from the argument for the preservation of the woods.

*Source*: **A**. John Wesley Powell, "The Irrigable Lands of the Arid Region," *Century Magazine* 39, no. 5 (March 1890): 766–68. **B**. John Wesley Powell, "The Non-irrigable Lands of the Arid Region," *Century Magazine* 39, no. 6 (April 1890): 917–20.

# Part IV

# The Conservation Movement Era, 1890–1920

As a major producer of iron and steel, coal and coke, oil, cotton and woolen goods, farm implements, and refined sugar, the United States in 1890 was a modern industrial powerhouse. It was also the fourth most populous country in the world, with a population of nearly 63 million. The steam engine and steam-powered tractors had already transformed farm life, especially on the large wheat farms of the Midwest, but few people had electricity or owned cars yet.

Over the next three decades, the nation's population would expand greatly as a result of massive immigration from southern and eastern Europe. Simultaneously, the number of households that possessed electric power and lighting, telephones, and/or automobiles would also increase tremendously. By 1920, immigration and technological innovation had altered the face of America.

## PUBLIC HEALTH AND SANITATION

As a result of scientific discoveries made in the second half of the nineteenth century, several major sanitary and medical advances—including the pasteurization of milk, the chlorination and filtration of water, and the use of antiseptics and other bacteria-destroying agents—were gradually introduced throughout much of the United States. Many cities had begun building both water supply and sewage systems around the 1850s, and by the turn of the century, greater cleanliness of water and food as well as improvements in medical care had led to an increase in life expectancy. While life expectancy in 1850 was only about 38.3 years for men and 40.5 years for women, by 1900 it had increased to 46.3 years for men and 48.3 years for women, and by 1920 it had risen to 53.6 years for men and 54.6 years for women.[1]

Nevertheless, pollution and poor sanitation remained problems in both urban and rural areas, in part because local efforts to stop pollution were constantly being challenged by industry. The disposal of garbage and waste, including the droppings of horses used to pull carriages, and the general need for better sanitation continued to endanger public health. Furthermore, the problems of industrial waste and soot that had plagued towns and cities in the nineteenth century [see Document 41] had intensified as industrialization increased and cities grew [see Document 67]. In reaction, a movement to improve health, living, and working conditions, especially in the cities, gained ground. It was spearheaded primarily by women, many of them advocates for the poor and working class (like Jane Addams [see Document 72] and Lillian Wald) and doctors, nurses, and scientists (like Alice Hamilton and Ellen Swallow Richards [see Document 65]).

Because pollution and unsanitary conditions were generally considered to be local issues, most social and legislative efforts to ameliorate these problems were initiated by local activists or members of state or local governments. The federal government evinced little interest in pollution and sanitation problems until 1899, when the Rivers and Harbors Act [see Document 61]—the first federal antipollution law—was passed.

While industrial growth and urban expansion fueled many urban sanitary problems, the movement of people away from the sources of food production also created new types of sanitary problems, related to food production and packaging, that called for federal oversight. This need for greater federal involvement together with the general demand for improved sanitation helped to sparked a pure foods movement. The movement, which was reinforced by the graphic descriptions of unsanitary conditions in the meat-packing industry in Upton Sinclair's novel *The Jungle* [see Document 64], provided the impetus for the passage of the Pure Food and Drug Act of 1906 and the Meat Inspection Act of 1907.

## THE FEDERAL CONSERVATION EFFORT

The recognition that there was no longer any frontier available in which to expand [see Document 60] stimulated increasing numbers of people to be concerned about the careless destruction and plunder of the nation's resources. In 1891 President Benjamin Harrison created a national forest system with a set-aside of 13 million acres of forested land [see Document 59]. The provision authorizing that the land be set aside was made in an obscure section of a bill concerned with the repeal of timber culture laws that is now known as the Forest Reserve Act. During President Grover Cleveland's second term of office (1893–1897), Harrison's forest reserve was augmented by another 21 million acres.

In 1901, twenty years after John Wesley Powell began his campaign to reclaim the arid lands of the West, Powell's cause was embraced by Teddy Roosevelt [see Document 62]. Then, in 1902, following the passage of the Reclamation Act [see Document 63], the federal government sold off massive amounts of public lands to provide funds for irrigation projects. Roosevelt Dam, on the Salt River in Arizona, the first of the large dams to be built with these funds, was completed in 1911. Roosevelt also embraced Powell's proposal to conserve the woodlands of the arid regions, for it was becoming obvious that the nation's woodlands were fast disappearing. By 1900 all but one-quarter of the virgin timberlands that had existed when the first colonists arrived had been cut down.

By 1907 President Theodore Roosevelt was deeply committed to the cause of resource conservation [see Document 66]. The first real effort of the federal government to develop a conservation policy began that year, when Roosevelt, at the urging of WJ McGee [see Document 71] and Gifford Pinchot [see Document 73], formed the Inland Waterways Commission to prepare a comprehensive national plan regarding flood control, irrigation, water transportation, hydroelectric development, and soil conservation. The commission, which included both McGee and Pinchot, determined that the issue of waterways and forest cover was of concern to the nation as a whole and recommended that the president convene all the state governors to discuss the problem. Invited guests at the conference, held at the White House on May 13, 1908, included congressmen and other prominent politicians, scientists, and Supreme Court justices in addition to the state governors. During the conference it became evident that very little was known about the true extent of available natural resources in the country, although it was obvious that a resource problem did exist. The National Conservation Commission [see Document 70] was created to look into the problem, but Congress failed to appropriate sufficient funds for the planned study.

Roosevelt, an ardent antimonopolist, also attempted to prevent the bleeding of America's resources by instituting antitrust legislation aimed at the huge monopolies that at the turn of the century controlled the railroads and the production of beef, sugar, fertilizers, and farm machinery.

In 1909 Roosevelt invited representatives from Canada, Newfoundland, and Mexico to a North American conservation conference. He also tried to organize a gathering of representatives from countries around the world at the Hague, but the proposal sparked little interest because most of the resource issues on his suggested agenda had already been dealt with by European nations.

Much of the government's "conservation" effort during the administration of Theodore Roosevelt and for a decade after was focused on land reclamation and the building of dams and reservoirs to supply the growing population of the West, where water was a problem because a large part of the region was arid or semiarid. A proposal to flood the magnificent Hetch Hetchy Valley led to a fierce battle between preservationists, such as John Muir, and wise-use conservationists, such as Teddy Roosevelt and Pinchot, that underlined a serious rift in the rapidly growing conservation movement [see Document 68]. Although the utilitarians—the wise-use advocates—won, the fight brought the preservationists together and galvanized the conservation movement.

In the ensuing years, Roosevelt's conservation initiatives continued to bear fruit. Discussions at the 1909 North American Conservation Conference had set the groundwork for multilateral agreements on migratory species, and in 1916 the United States and Canada signed a Migratory Bird Treaty.

## THE CHANGING LANDSCAPE

Since the birth of the nation, human activity and population growth had been changing the natural landscape, but in the twentieth century, as the rate of change increased markedly, many species of plants and animals reached the brink of extinction. By 1914, the passenger pigeon, which had once darkened the sky and provided food for thousands of people, was gone, and oyster supplies for New York City markets had dried up as result of pollution in the city's harbor.

In 1917, though, the attention of the United States was focused on fighting World War I. Efforts were made to fortify the shipbuilding industry and to develop the infant aviation industry. As a result of wartime demands, large numbers of people found jobs in the industrial sector and moved to the cities.

## DOCUMENT 59: Forest Reserve Act (1891)

The clause establishing the U.S. Forest Reserve System was buried, as section 24, in a bill to repeal timber culture laws. Following the passage of the bill, President Benjamin Harrison set aside 13 million acres of forest land.

[T]he President of the United States may, from time to time, set apart and reserve, in any State or Territory having public land bearing forests, in any part of the public lands wholly or in part covered with timber or undergrowth, whether of commercial value or not, as public reservations, and the President shall, by public proclamation, declare the establishment of such reservations and the limits thereof.

*Source: United States Statutes at Large,* Vol. 26 (Washington, D.C.: Government Printing Office, 1892), 51st Cong., 2nd sess., chap. 561, March 3, 1891, p. 1103.

---

## DOCUMENT 60: Frederick J. Turner on the Disappearance of the Frontier (1894)

---

Until the late nineteenth century, the existence of a continuously advancing area that contained free land and undiscovered resources was a primary factor in the development of the United States and in the thought and actions of the nation's inhabitants. Once the frontier reached the Pacific Ocean, Americans were forced to confront such issues as resource limitations and land availability for an ever increasing population. The Harvard professor of history Frederick Jackson Turner was prompted to write his definitive essay on the closing of the American frontier by the Census Report of 1890, which stated that the "frontier of settlement" no longer existed and that future census reports would not include data on the frontier.[2]

The exploitation of the beasts took hunter and trader to the west, the exploitation of the grasses took the rancher west, and the exploitation of the virgin soil of the river valleys and prairies attracted the farmer. Good soils have been the most continuous attraction to the farmer's frontier. The land hunger of the Virginians drew them down the rivers into Carolina, in the early colonial days; the search for soils took the Massachusetts men to Pennsylvania and to New York. As the eastern lands were taken up migration flowed across them to the west. Daniel Boone, the great backwoodsman, who combined the occupations of hunter, trader, cattle-raiser, farmer, and surveyor—learning, probably from the traders, of the fertility of the lands on the upper Yadkin, where the traders were wont to rest as they took their way to the Indians, left his Pennsylvania home with his father, and passed down the Great Valley road to that stream. Learning from a trader whose posts were on the Red River in Kentucky of its game and rich pastures, he pioneered the way for the farmers to that region. Thence he passed to the frontier of Missouri, where his settlement was long a landmark on the frontier. Here again he helped to open the way for civilization, finding salt licks, and trails, and land. His son was among the earliest trappers in the passes of the Rocky Mountains.

* * *

Obviously the immigrant was attracted by the cheap lands of the frontier, and even the native farmer felt their influence strongly. Year by year the farmers who lived on soil whose returns were diminished by unrotated crops were offered the virgin soil of the frontier at nominal prices. Their growing families demanded more lands, and these were dear. The competition of the unexhausted, cheap, and easily tilled prairie lands compelled the farmer either to go west and continue the exhaustion of the soil on a new frontier, or to adopt intensive culture. . . . Thus the demand for land and the love of wilderness freedom drew the frontier ever onward.

* * *

The stubborn American environment is [at the frontier] with its imperious summons to accept its conditions; the inherited ways of doing things are also there; and yet, in spite of environment, and in spite of custom, each frontier did indeed furnish a new field of opportunity, a gate of escape from the bondage of the past; and freshness, and confidence, and scorn of older society, impatience of its restraints and its ideas, and indifference to its lessons, have accompanied the frontier. What the Mediterranean Sea was to the Greeks, breaking the bond of custom, offering new experiences, calling out new institutions and activities, that, and more, the ever retreating frontier has been to the United States directly, and to the nations of Europe more remotely. And now, four centuries from the discovery of America, at the end of a hundred years of life under the Constitution, the frontier has gone, and with its going has closed the first period of American history.

Source: Frederick J. Turner, The Significance of the Frontier in American History (Washington, D.C.: Government Printing Office, 1894; Readex Facsimile edition), pp. 213, 215, 227.

## DOCUMENT 61: Rivers and Harbors Act (1899)

By the end of the nineteenth century industrial wastes were beginning to have a deleterious effect on the water quality of rivers around the country. Tucked away in section 13 of the Rivers and Harbors Act of 1899—which was primarily an appropriations bill allocating funds for dozens of river and harbor construction, repair, and preservation projects—was a prohibition against dumping refuse in navigable waters, making it the first federal antipollution law. It is interesting to note that the first colonial antipollution statute also concerned the dumping of garbage in harbors [see Document 12].

[I]t shall not be lawful to throw, discharge, or deposit, or cause, suffer, or procure to be thrown, discharged, or deposited either from or out of any ship, barge, or other floating craft of any kind, or from the shore, wharf, manufacturing establishment, or mill of any kind, any refuse matter of any kind or description whatever other than that flowing from streets and sewers and passing therefrom in a liquid state, into any navigable water of the United States, or into any tributary of any navigable water from which the same shall float or be washed into such navigable water; and it shall not be lawful to deposit, or cause, suffer, or procure to be deposited material of any kind in any place on the bank of any navigable water, or on the bank of any tributary of any navigable water, where the same shall be liable to be washed into such navigable water; either by ordinary or high tides, or by storms or floods, or otherwise, whereby navigation shall or may be impeded or obstructed.

*Source: U.S. Statutes at Large*, Vol. 30, Part II (Washington, D.C.: Government Printing Office, 1900), 55th Cong., 3rd sess., chap. 425, March 3, 1899, p. 1152.

---

## DOCUMENT 62: Theodore Roosevelt on Forest Conservation and Land Reclamation (1901)

Teddy Roosevelt knew and loved the wild areas of the country, but at heart he was a pragmatic utilitarian whose conservation policies were driven by a desire to manage the land so that it would produce more game to hunt, more trees for timber, more water for irrigation, and more land for farming. Much of the content of this address, which set the stage for the Reclamation Act [see Document 63], is a reworking of John Wesley Powell's Reports on the Lands of the Arid Region [see Document 58]. Some of Roosevelt's later conservation policies were influenced by John Muir [Document 68] and other preservationists, who convinced him of the value of actually setting aside wilderness areas.

Public opinion throughout the United States has moved steadily toward a just appreciation of the nature of forests, whether of planted or natural growth. The great part played by them in the creation and maintenance of the National wealth is now more fully realized than ever before.

Wise forest protection does not mean the withdrawal of forest resources, whether of wood, water, or grass, from contributing their full share to the welfare of the people, but, on the contrary, gives the assurance of larger and more certain supplies. The fundamental idea of forestry is the perpetuation of forests by use. Forest protection is not an end of itself. It is a means to increase and sustain the resources of our country and the industries which depend upon them. The preservation of our forests is an imperative business necessity. We have come to see clearly that whatsoever destroys the forest, except to make way for agriculture, threatens our well-being.

The practical usefulness of the National forest reserves to the mining, grazing, irrigation, and other interests of the regions in which the reserves lie has led to widespread demand by the people of the West for their protection and extension. The forest reserves will inevitably be of still greater use in the future than in the past. Additions should be made to them whenever practicable, and their usefulness should be increased by a thoroughly businesslike management.

\* \* \*

The wise Administration of the forest reserves will be not less useful to the interests which depend on water than to those which depend on wood and grazing. The water supply itself depends upon the forest. In the arid region it is water, not land, which measures production. The western half of the United States would sustain a population greater than that of our whole country today if the waters that now run to waste were saved and used for irrigation. The forest and water problems are perhaps the most vital internal questions of the United States.

\* \* \*

The forest alone cannot, however, fully regulate and conserve the waters of the arid regions. Great storage works are necessary to equalize the flow of streams and so save the flood waters. Their construction has been shown to be an undertaking too vast for private effort. Nor can it be best accomplished by the individual States working alone. Far-reaching inter-State problems are involved and the resources of individual States would often be inadequate. It is properly a National function, at least in some of its features. It is as right for the National Government to make the streams and rivers of the arid region useful by engineering works for water storage as to make useful the rivers and harbors of the humid region by engineering works of another kind. The storing of the floods in reservoirs at the headwaters of our rivers is but an enlargement of our present policy of river control, under which levees are built at the lower reaches of the same streams.

* * *

Our aim should be not simply to reclaim the largest areas of land and provide homes for the largest numbers of people, but to create for this new industry the best possible social and industrial conditions.

Source: Theodore Roosevelt, "Address to Congress," December 3, 1901, New York Times, December 4, 1901, p. 4.

## DOCUMENT 63: Reclamation Act (1902)

In 1902 John Wesley Powell's grand plan for the arid lands of the United States [see Document 58], after languishing for thirty-four years, finally became federal policy. Although in previous centuries industrious farmers had watered their dry lands in various parts of the country, prior to the implementation of the Reclamation Act irrigation had never been carried out on a large scale in the United States. As a result of the Reclamation Act, new lands were opened for settlement and farming, and in the process the precarious ecological balance of vast areas of our country was upset. Three-quarters of a century after the act's passage, people began to question the wisdom of the act and to consider undoing some of the follies that stemmed from it [see Documents 121 and 131].

Be it enacted . . . , That all moneys received from the sale and disposal of public lands in Arizona, California, Colorado, Idaho, Kansas, Montana, Nebraska, Nevada, New Mexico, North Dakota, Oklahoma, Oregon, Utah, Washington, and Wyoming, beginning with the fiscal year ending June thirtieth, nineteen hundred and one, including the surplus of fees and commissions in excess of allowances to registers and receivers, and excepting the five per centum of the proceeds of the sales of public lands in the above States set aside by law for educational and other purposes, shall be, and the same are hereby, reserved, set aside, and appropriated as a special fund in the Treasury to be known as the "reclamation fund," to be used in the examination and survey for and the construction and maintenance of irrigation works for the storage, diversion, and development of waters for the reclamation of arid and semiarid lands in the said States and Territories, and for the payment of all other expenditures provided for in this Act.

*Source: United States Statutes at Large,* Vol. 32, Part I (Washington, D.C.: Government Printing Office, 1903), 57th Cong., 1st sess., chap. 1093, June 17, 1902, p. 388.

## DOCUMENT 64: Upton Sinclair on the Adulteration of Processed Food (1906)

Upton Sinclair's novel *The Jungle* was a fictionalized exposé of abuses in the meat-packing industry. It raised the public's consciousness about such issues as the lack of quality control in the food industry, the mis-labeling of packaged foods and other packaged goods, the deliberate use of adulterated products by the packaged foods industry, and the inhumane treatment of animals by slaughterhouses.

It was only when the whole ham was spoiled that it came into the department of Elzbieta. Cut up by the two-thousand-revolutions-a-minute flyers, and mixed with half a ton of other meat, no odor that ever was in a ham could make any difference. There was never the least attention paid to what was cut up for sausage; there would come all the way back from Europe old sausage that had been rejected, and that was mouldy and white—it would be dosed with borax and glycerine, and dumped into the hoppers, and made over again for home consumption. There would be meat that had tumbled out on the floor, in the dirt and sawdust, where the workers had tramped and spit uncounted billions of consumption germs. There would be meat stored in great piles in rooms; and the water from leaky roofs would drip over it, and thousands of rats would race about on it. It was too dark in these storage places to see well, but a man could run his hand over these piles of meat and sweep off handfuls of the dried dung of rats. These rats were nuisances, and the packers would put poisoned bread out for them, they would die, and then rats, bread, and meat would go into the hoppers together. ... There was no place for the men to wash their hands before they ate their dinner, and so they made a practice of washing them in the water that was to be ladled into the sausage. There were the butt-ends of smoked meat, and the scraps of corned beef, and all the odds and ends of the waste of the plants, that would be dumped into old barrels in the cellar and left there. Under the system of rigid economy which the pack-ers enforced, there were some jobs that it only paid to do once in a long time, and among these was the cleaning out of the waste barrels. Every spring they did it; and in the barrels would be dirt and rust and old nails and stale water—and cart load after cart load of it would be taken up and dumped into the hoppers with fresh meat, and sent out to the public's breakfast. Some of it they would make into "smoked" sausage—but as the smoking took time, and was therefore expensive, they would call upon their chemistry department, and preserve it with borax and color it with gelatine to make it brown. All of their sausage came out of the same bowl, but when they came to wrap it they would stamp some of it "special," and for this they would charge two cents more a pound.

*Source*: Upton Sinclair, *The Jungle* (New York: Doubleday, Page, 1906), pp. 161–62.

## DOCUMENT 65: Ellen Swallow Richards on Sanitation and Human Ecology (1907)

Ellen Swallow Richards, an instructor in sanitary chemistry at Massachusetts Institute of Technology, is credited with introducing the term *human ecology* and with popularizing the concept of ecology and the idea that people should be concerned about and take responsibility for their environment.

Sanitary science teaches that mode of life which promotes health and efficiency.

The individual is one of a community influencing and influenced by the common environment.

Human ecology is the study of the surroundings of human beings in the effects they produce on the lives of men. The features of the environment are natural, as climate, and artificial, produced by human activity, as noise, dust, poisonous vapors, vitiated air, dirty water, and unclean food.

The study of this environment is in two chief lines:

First, what is often called municipal housekeeping—the co-operation of the citizens in securing clean streets, the suppression of nuisances, abundant water supply, market inspection, etc.

Second, family housekeeping. The healthful home demands a management of the house which shall promote vigorous life and prevent the physical deterioration so evident under modern conditions.

* * *

To secure and maintain a safe environment there must be inculcated *habits* of using the material things in daily life in such a way as to promote and not to diminish health. Avoid spitting in the streets, avoid throwing refuse on the sidewalk, avoid dust and bad air in the house and sleeping room, etc.

It is, however, of the greatest importance that every one should acquire such habits of *belief* in the importance of this material environment as shall lead him to insist upon sanitary regulations, and to see that they are carried out.

What touches my neighbor, touches me. For my sake, and for his, the city inspector and the city garbage cart visit us, and I keep my premises in such a condition as I expect him to strive for.

The first law of sanitation requires quick removal and destruction of all wastes—of things done with.

The second law enjoins such use of the air, water, and food necessary to life that the person may be in a state of health and efficiency.

This right use depends so largely upon habit that a great portion of sanitary teaching must be given to inculcating right and safe ways in daily life.

*Source*: Ellen Swallow Richards, *Sanitation in Daily Life* (Boston: Whitcomb & Barrows, 1910 [1907]), pp. v–viii, in Carolyn Merchant, ed., *Major Problems in American Environmental History: Documents and Essays* (Lexington, MA: Heath, 1993), pp. 445–47.

---

## DOCUMENT 66: Theodore Roosevelt on the Conservation and Use of Natural Resources (1907)

In his December 3, 1907, annual address to Congress, President Roosevelt focused on the need to use the resources of the nation prudently, and he attacked those who were willing to exhaust the nation's resources in the process of fattening their own pockets. As an arch opponent of the concentration of extensive power in the hands of big business, Roosevelt fought the environmental depredations of business conglomerates as well as of ranching and other special interest groups.

The conservation of our natural resources and their proper use constitute the fundamental problem which underlies almost every other problem of our National life. We must maintain for our civilization the adequate material basis without which that civilization can not exist. We must show foresight, we must look ahead. As a nation we not only enjoy a wonderful measure of present prosperity but if this prosperity is used aright it is an earnest of success such as no other nation will have. The reward of foresight for this Nation is great and easily foretold. But there must be the look ahead, there must be a realization of the fact that to waste, to destroy, our natural resources, to skin and exhaust land instead of using it so as to increase its usefulness, will result in undermining in the days of our children the very prosperity which we ought by right to hand down to them amplified and developed. For the last few years, through several agencies, the Government has been endeavoring to get our people to look ahead and to substitute a planned and orderly development of our resources in place of a haphazard striving for immediate profit. . . .

Irrigation should be far more extensively developed than at present, not only in the States of the Great Plains and the Rocky Mountains, but in many others, as, for instance, in large portions of the South Atlantic and Gulf States, where it should go hand in hand with the reclamation of swamp land. The Federal Government should seriously devote itself to this task, realizing that utilization of waterways and water-power, forestry, irrigation, and the reclamation of lands threatened with overflow, are all interdependent parts of the same problem. The work of the Reclamation Service in developing the larger opportunities of the western half of our country for irrigation is more important than almost any other movement. The constant purpose of the Government in connection with the Reclamation Service has been to use the water resources of the public lands for the ultimate greatest good of the greatest number; in other words, to put upon the land permanent home-makers, to use and develop it for themselves and for their children and children's children. There has been, of course, opposition to this work; opposition from some interested men who desire to exhaust the land for their own immediate profit without regard to the welfare of the next generation, and opposition from honest and well-meaning men who did not fully understand the subject or who did not look far enough ahead.

* * *

Some such legislation as that proposed [by the Public Lands Commission] is essential in order to preserve the great stretches of public grazing land which are unfit for agriculture under present methods. . . . As the West settles the range becomes more and more over-grazed. Much of it can not be used to advantage unless it is fenced, for fencing is the only way by which to keep in check the owners of nomad flock which roam hither and thither, utterly destroying the pastures. . . .

. . . . We are prone to speak of the resources of this country as inexhaustible; this is not so. The mineral wealth of the country, the coal, iron, oil, gas, and the like, does not reproduce itself, and therefore is certain to be exhausted ultimately; and wastefulness in dealing with it to-day means that our descendants will feel the exhaustion a generation or two before they otherwise would.

*Source*: Theodore Roosevelt, address to Congress, in *Congressional Record—Senate*, December 3, 1907 (Washington, D.C.: Government Printing Office, 1907), pp. 74–76.

---

## DOCUMENT 67: Frederick Law Olmsted, Jr., on the Smoke Nuisance (1908)

---

Frederick Law Olmsted, Jr., whose given name was actually Henry Perkins, was the son of Frederick Law Olmsted [see Document 48]. As a young man he worked for his father's landscape architecture firm. Not until nearly half a century after the publication of this article did people begin to comprehend the full range of the impacts of soot (particulate matter) on the health of both humans and their environment.

The dweller in a town burning bituminous coal needs no definition of the smoke nuisance. The great cloud that hangs over the city like a pall can be seen from any neighboring hilltop, and the dweller within is only too well aware of the splotches of soot that settle on every object in the city, bedimming buildings, spoiling curtains, injuring books, and increasing the laundry bill. The direct menace to the public health in fostering tuberculous conditions by loading the air with carbon particles to lodge in the lungs, and by causing housekeepers to keep the windows shut for fear of the soot that floats in when they are open, is equaled only by the mentally and physically depressing effect of the pall which shuts out the life-giving and germ-destroying sunshine. Our city parks have mostly lost their evergreen character, where it existed, as conifers cannot long endure city smoke. Thus one treatment of the most pleasing variations in landscape is made impossible.

\* \* \*

*There should be complete understanding of the scientific fact that visible black smoke is made up almost entirely of unconsumed particles of combustible carbon, or coal, wasted into the atmosphere through imperfect combustion. It is economic waste, in itself; and its emission creates additional waste.*

*No really intelligent person now denies the imperative economic and sanitary need for abating or suppressing the smoke evil, nor the feasibility and absolute power of existing authorities to do so where the will and proper public sentiment exist.*

The tearing down of a dangerous house, the draining of a pestiferous swamp, the cleaning of a filthy street, or of a back yard, are simple remedies for simple nuisances. The abolition of smoke, on the other hand, affects the whole community, since the production of smoke is claimed, especially by the careless or the uninformed, to be completely bound up with the material and industrial welfare of a city. The evil is one that grows with the growth of the community, and its abatement calls for a large, comprehensive and tactful treatment, with thorough cooperation between the different parties to the problem. Education of the public, the factory owners and the firemen to the bad economy and the wrong of smoke emission is of great importance.

\* \* \*

The first step in abating smoke is to pass a law or an ordinance, making the emission of black or dark gray smoke an unlawful act, punishable by fine. Such regulations are already in force in New York, Cleveland, Milwaukee, Toronto, Toledo, Indianapolis, Detroit, the District of Columbia, and numerous other cities. The second step is to get the law enforced.

\* \* \*

In every case, *smoke is a preventable nuisance*, and every smoking plant or locomotive is a sign of wastefulness, and a disregard for the rights of the public. The proprietor should be as interested in abating the nuisance as his neighbors, and it has been the experience of smoke-law officials that men who have bitterly complained at being forced to make improvements have afterward thanked the smoke-abating department for the increased economy of the plant.

*Source*: Frederick Law Olmsted et al., *The Smoke Nuisance*, American Civic Association (Philadelphia) Series II, no. 1 (March 1908), pp. 4–6.

## Document 68: John Muir, James Phelan, and the Battle over the Flooding of the Hetch Hetchy Valley (1908–1913)

John Muir was one of the most influential conservationists in the last decades of the nineteenth century and the first decades of the twentieth century. In 1889 he became a major force advocating the development of Yosemite as a national park, and in 1892 he helped found the Sierra Club, the first western-based hiking society. A three-day camping trip that President Theodore Roosevelt took with Muir in the Sierra Nevada Mountains of eastern California in 1903 made a deep impression on Roosevelt and was influential in shaping government policies concerning wilderness and wildlife. Many believe that Roosevelt set aside 148 million acres of forest reserve land, in addition to establishing the first National Wildlife Refuge in 1903 and pioneering the federal role in conservation, partly as a result of his relationship with Muir.

When it was proposed that the Hetch Hetchy Valley be flooded to build a reservoir to supply water for the city of San Francisco, Muir launched a fight to save the valley. Although the federal government gave the Hetch Hetchy to San Francisco in 1913, the Muir-led struggle galvanized the nascent conservation movement.

James Phelan was mayor of San Francisco from 1897 to 1902.

## A. John Muir to Theodore Roosevelt, April 21, 1908

I am anxious that the Yosemite National Park may be saved from all sorts of commercialism and marks of man's work other than the roads, hotels, etc., required to make its wonders and blessings available. For as far as I have seen there is not in all the wonderful Sierra, or indeed in the world, another so grand and wonderful block of Nature's mountain handiwork.

There is now under consideration, as doubtless you well know, an application of San Francisco supervisors for the use of the Hetch-Hetchy Valley and Lake Eleanor as storage reservoirs for a city water supply. This application should, I think, be denied, especially the Hetch-Hetchy part, for this Valley . . . is a counterpart of Yosemite, and one of the most sublime and beautiful and important features of the Park.

## B. James Phelan, Letter to *Outlook*, 1909

The Hetch-Hetchy is one of a dozen mountain gorges, and, while beautiful, it is not unique. It is accessible over difficult trails about three months during the year, and few ever visit it. The Yosemite Valley satisfies every craving for large numbers of tourists, and the State of California, a few years ago freely ceded this Valley to the Federal Government, and at the same time purchased a great redwood forest in the interest of forest preservation. California would not countenance the desecration of any of her scenery, and yet the State Legislature, now in session, has unanimously petitioned Congress to pass this bill. President Roosevelt, Secretary of the Interior Garfield, Forester Pinchot, will yield to none in their love of nature; yet they strongly favor this bill. . . . The only question is, after all, the conversion of the Hetch-Hetchy Meadow into a crystal clear Lake—a natural object of indeed rare beauty. . . .

By yielding their opposition, sincere lovers of nature will turn the prayers of a million people to praise for the gifts bestowed upon them by the God of Nature, whom they cannot worship in his temple, but must perforce live in the sweltering cities. A reduced death rate is a more vital consideration than the discussion of the relative beauties of a meadow or a lake.

### C. From John Muir's *The Yosemite*, 1912

Hetch Hetchy, they say, is a "low-lying meadow." On the contrary, it is a high-lying natural landscape garden. . . .

"It is a common minor feature, like thousands of others." On the contrary it is a very uncommon feature; after Yosemite, the rarest and in many ways the most important in the National Park.

"Damming and submerging it 175 feet deep would enhance its beauty by forming a crystal-clear lake." Landscape gardens, places of recreation and worship, are never made beautiful by destroying and burying them. The beautiful sham lake, forsooth, would be only an eyesore, a dismal blot on the landscape, like many others to be seen in the Sierra. For, instead of keeping it at the same level all the year, allowing Nature centuries of time to make new shores, it would, of course, be full only a month or two in the spring, when the snow is melting fast; then it would be gradually drained, exposing the slimy sides of the basin and shallower parts of the bottom, with the gathered drift and waste, death and decay of the upper basins, caught here instead of being swept on to decent natural burial along the banks of the river or in the sea. Thus the Hetch Hetchy dam-lake would be only a rough imitation of a natural lake for a few of the spring months, an open sepulcher for the others.

"Hetch Hetchy water is the purest of all to be found in the Sierra, unpolluted, and forever unpollutable." On the contrary, excepting that of the Merced below Yosemite, it is less pure than that of most of the other Sierra streams, because of the sewerage of camp grounds draining into it, especially of the Big Tuolumne Meadows camp ground, occupied by hundreds of tourists and mountaineers, with their animals, for months every summer, soon to be followed by thousands from all the world.

These temple destroyers, devotees of ravaging commercialism, seem to have a perfect contempt for Nature, and, instead of lifting their eyes to the God of the mountains, lift them to the Almighty dollar.

Dam Hetch Hetchy! As well dam for water-tanks the people's cathedrals and churches, for no holier temple has ever been consecrated by the heart of man.

*Source*: **A.** John Muir to Theodore Roosevelt, in William Frederic Badé, ed., *The Life and Letters of John Muir*, Vol. 2 (Cambridge, MA, 1924), quoted in Robert McHenry and Charles Van Doren, eds., *A Documentary History of Conservation in* *the United States* (New York: Praeger, 1972), p. 307. **B**. James D. Phelan, "Dam Hetch-Hetchy," letter to *Outlook*, February 13, 1909, in McHenry and Van Doren, *Documentary History*, pp. 309–10. **C**. John Muir, *The Yosemite* (Madison: University of Wisconsin Press, 1986; reprint of 1912 Century ed.), pp. 260–62.

## DOCUMENT 69: Richard Ballinger on the Development of the West (1909)

President William Howard Taft's secretary of the Interior, Richard Ballinger, was a fierce opponent of the proposals made by Pinchot, Powell, and other conservationists for government oversight of the development of publicly owned lands containing forest, water, and mineral resources. His advocacy of corporate control of land and resource development was echoed nearly three-quarters of a century later when James Watt became secretary of the Interior under Ronald Reagan.

### A. From an Interview with John L. Mathews

MATHEWS: What is your object in giving in to the railroads and letting them destroy this water power [along the Deschutes River in Oregon]?

BALLINGER: . . . You chaps who are in favor of this Conservation programme are all wrong. You are hindering the development of the West. These railroads are necessary to the country. And more than that, this whole big [public] domain is a blanket—it is oppressing the people. The thing to do with it—In my opinion, the proper course to take with regard to this domain is to divide it up among the big corporations and the people who know how to make money out of it and let the people at large get the benefits of the circulation of the money.

### B. From Address to the National Irrigation Congress, August 12, 1909

While the Government has invested over fifty million dollars in irrigation works, many times that amount has been invested since the passage of the Reclamation Act by private enterprise and it is safe to say that a large portion of these private investments have resulted from Governmental example and encouragement; and let me say here that it has not been and is not the policy of the National Government in the administration of this act to hinder or interfere with the investment of private capital in the construction of irrigation works but rather to lend it encouragement. This is particularly true in reference to irrigation under the Carey Act [of 1894] in the various States.

I am not a believer in the Government entering into competition with legitimate private enterprise. Its functions under the Reclamation Act are not of this character. . . .

The purpose of the Reclamation Act is to undertake the irrigation of arid and semi-arid lands where a considerable portion thereof belongs to the public domain, and by the installation of the storage and diversion of available waters to irrigate the largest possible area within a given territory at the least cost to the entrymen and land owners for construction, maintenance and operation, always keeping in view the matter of the settlement of these lands and rendering them capable of supporting the greatest number of families. While it is a reclamation act, it is also a settlement act, and the public lands which are proposed to be irrigated by means of the contemplated works have been rendered subject to entry only under the homestead laws in small tracts capable of supporting a family. . . . [O]nly the cost of construction and maintenance shall be repaid to the Government.

. . .

Any one who has visited one or more of the Reclamation projects now in operation and sees on the one hand the desert covered with sage brush and barrenness, and on the other, the water flowing over the fertile soil producing heavy crops of grain, or orchards in fruit, appreciates to the fullest extent the benefits of irrigation.

The people of the West, therefore, who are familiar with these wonderful results in irrigation, are highly appreciative of the importance of the Reclamation Service, but the great difficulty which that service encounters is in finishing the projects now undertaken as against the clamor for a diversion of the funds to new fields.

. . .

The danger, which the Government is undertaking to overcome, is the establishment of small irrigation projects in localities where by such establishment the larger opportunities are destroyed, thus preventing enormous areas of lands from ever acquiring the use of water. It is quite true that many small projects capable of being financed by men of limited means can be carved out of larger possibilities, but to encourage them means the loss of the larger possibilities.

. . .

Much of [our nation's wealth] has been accumulated by the destruction, by the sacrifice and waste of nature's gifts, and it is a fortuitous circumstance that the country has been brought to understand the importance of utilizing and saving our natural wealth and making it possible for the nation to continue to prosper, and for the generations that are to come to have some share in that prosperity, especially since no element of the nation's wealth is greater than that contained in the soil. For this reason, if for no other, the work of reclamation of the arid and semi-arid lands of the West is worthy of first importance in the development of the nation's resources.

*Source*: **A**. Richard Ballinger, interview with John L. Mathews, Washington, D.C., quoted in John L. Mathews, "Mr. Ballinger and the National Grab-Bag," *Hampton's Magazine*, December 1909, in Richard A. Ballinger Papers, Manuscript Collection of the University of Washington, Seattle, microfilm roll 12. **B**. R. A. Ballinger, "Attitude of the Administration toward the Reclamation of the Arid Lands in the West," remarks made at the National Irrigation Congress, Spokane, WA, August 12, 1909, in Richard A. Ballinger Papers, microfilm roll 11.

## DOCUMENT 70: Report of the National Conservation Commission (1909)

As a result of the governor's conference on conservation, convened by Roosevelt in 1908, the National Conservation Commission was appointed. The chairman of the commission was Gifford Pinchot [see Document 73]. Unfortunately, Congress failed to allocate adequate funding for the commission to make a thorough study of the nation's resources. That study had to wait until 1952 and the appointment of the President's Materials Policy Commission on Economic Growth and Resource Policy [see Document 89]. This selection from the National Conservation Commission report emphasizes the development of a program of "wise and beneficial uses" of natural resources.

In this country, blessed with natural resources in unsurpassed profusion, the sense of responsibility to the future has been slow to awaken. Beginning without appreciation of the measure or the value of natural resources other than land with water for commercial uses, our forefathers pushed into the wilderness and, through a spirit of enterprise which is the glory of the nation, developed other great resources. Forests were cleared away as obstacles to the use of the land; iron and coal were discovered and developed, though for years their presence added nothing to the price of the land; and through the use of native woods and metals and fuels, manufacturing grew beyond all precedent, and the country became a power among the nations of the world.

Gradually the timber growing on the ground and the iron and coal within the ground came to have a market value and were bought and sold as sources of wealth. Meanwhile, vast holdings of these resources were acquired by those of greater foresight than their neighbors before it was generally realized that they possessed value in themselves; and in this way large interests, assuming monopolistic proportions, grew up, with greater enrichment to their holders than the world had seen before, and with the motive of immediate profit, with no concern for the future or thought of the permanent benefit of country and people, a wasteful and profligate use of the resources began and has continued.

The waters, at first recognized only as aids to commerce in supplying transportation routes, were largely neglected. In time this neglect began to be noticed, and along with it the destruction and approaching exhaustion of the forests. This, in turn, directed attention to the rapid depletion of the coal and iron deposits and the misuse of the land.

\* \* \*

In the stage which we are entering wise and beneficial uses are essential, and the checking of waste is absolutely demanded.

\* \* \*

The wastes which most urgently require checking vary widely in character and amount. The most reprehensible waste is that of destruction, as in forest fires, uncontrolled flow of gas and oil, soil wash, and abandonment of coal in the mines. This is attributable, for the most part, to ignorance, indifference, or false notions of economy, to rectify which is the business of the people collectively.

Nearly as reprehensible is the waste arising from misuse, as in the consumption of fuel in furnaces and engines of low efficiency, the loss of water in floods, the employment of ill-adapted structural materials, the growing of ill-chosen crops, and the perpetuation of inferior stocks of plants and animals, all of which may be remedied.

*Source: Report of National Conservation Commission*, Vol. I, 60th Cong., 2nd sess., Senate Document 676 (Washington, D.C.: Government Printing Office, 1909), pp. 13–14.

---

## DOCUMENT 71: WJ McGee on Conservation (1909)

William John (always referred to as WJ, without periods, at his insistence) McGee became involved in federal resource work when he joined the United States Geological Survey in 1878, at the invitation of John Wesley Powell [see Document 58]. His interest in managing natural resources to serve the public was sparked at a meeting of the Lakes-to-the-Gulf Deep Waterway Association in 1906. This was several months before Gifford Pinchot became an active conservationist [see Document 73] and joined McGee in his effort to have the Inland Waterways Commission established by President Theodore Roosevelt. The appointment of this commission on March 14, 1907, marked the beginning of a national crusade in support of conservation.

[Gifford] Pinchot and [James] Garfield [secretary of the Interior, 1907–1909] especially, and [President Theodore] Roosevelt in his turn, sought to counteract the tendency toward wholesale alienation of the public lands for the benefit of the corporation and the oppression or suppression of the settler; and in the end their efforts resulted in what is now known as the Conservation Movement....

On its face the Conservation Movement is material—ultra-material.... Yet in truth there has never been in all human history a popular movement more firmly grounded in ethics, in the eternal equities, in the divinity of human rights! Whether we rise into the spiritual empyrean or cling more closely to the essence of humanity, we find our loftiest ideals made real in the Cult of Conservation....

... What *right* has any citizen of a free country, whatever his foresight and shrewdness, to seize on sources of life for his own behoof that are the common heritage of all; what *right* has legislature or court to help in the seizure; and striking still more deeply, what *right* has any generation to wholly consume, much less to waste, those sources of life without which the children or the children's children must starve or freeze? These are among the questions arising among intelligent minds in every part of this country, and giving form to a national feeling which is gradually rising to a new plane of equity. The questions will not down.... How shall they find answer? The ethical doctrine of Conservation answers: by a nobler patriotism, under which citizen-electors will cleave more strongly to their birthright of independence and strive more vigorously for purity of the ballot, for rightness in laws, for cleanness in courts, and for forthrightness in administration; by a higher honesty of purpose between man and man; by a warmer charity, under which the good of all will more fairly merge with the good of each; by a stronger family sense, tending toward a realization of the rights of the unborn; by deeper probity, maturing in the realizing sense that each holder of the sources of life is but a trustee for his nominal possessions, and is responsible to all men and for all time for making the best use of them in common interest; and by a livelier humanity, in which each will feel that he lives not for himself alone but as a part of a common life for a common world and for the common good....

... The American Revolution was fought for Liberty; the Constitution was framed for Equality; yet that third of the trinity of human impulses without which Union is not made perfect—Fraternity—has not been established: full brotherhood among men and generations has not yet come. The duty of the [Founding] Fathers was done well according to their lights; but some new light has come out of the West where their sons have striven against Nature's forces no less fiercely than the Fathers against foreign dominion. So it would seem to remain for Conservation to perfect the concept and the movement started among the Colonists one hundred and forty years ago—to round out the American Revolution by framing a clearer Bill of Rights. Whatever others there may be, surely these are inherent and indefeasible:—

1. The equal Rights of all men to opportunity.
2. The equal Rights of the People in and to resources rendered valuable by their own natural growth and orderly development.
3. The equal Rights of present and future generations in and to the resources of the country.
4. The equal Rights (and full responsibilities) of all citizens to provide for the perpetuity of families and States and the Union of States.

*Source*: WJ McGee, "The Conservation of Natural Resources," in *Proceedings of the Mississippi Valley Historical Association* 3 (1909–1910): 376–79, in Roderick Frazier Nash, ed., *Readings in the History of Conservation* (Reading, MA: Addison-Wesley, 1968), pp. 45–46.

## DOCUMENT 72: Jane Addams on Garbage (1910)

The Chicago social worker Jane Addams was one of the leading advocates of improved sanitation in the homes and the neighborhoods of the poor. Here she notes that the problem of garbage in lower-class neighborhoods was greater than in wealthier neighborhoods. Almost a hundred years later, this is still the case.

One of the striking features of our neighborhood twenty years ago, and one to which we never became reconciled, was the presence of huge wooden garbage boxes fastened to the street pavement in which the undisturbed refuse accumulated day by day. The system of garbage collecting was inadequate throughout the city but it became the greatest menace in a ward such as ours, where the normal amount of waste was much increased by the decayed fruit and vegetables discarded by the Italian and Greek fruit peddlers, and by the residuum left over from the piles of filthy rags which were fished out of the city dumps and brought to the homes of the rag pickers for further sorting and washing.

The children of the neighborhood twenty years ago played their games in and around these huge garbage boxes.

* * *

[My attempt to get a contract to remove garbage from the nineteenth ward] induced the mayor to appoint me the garbage inspector of the ward.

. . . The position was no sinecure whether regarded from the point of view of getting up at six in the morning to see that the men were early at work; or of following the loaded wagons, uneasily dropping their contents at intervals, to their dreary destination at the dump; or of insisting that the contractor must increase the number of his wagons from nine to thirteen and from thirteen to seventeen, although he assured me that he lost money on every one and that the former inspector had let him off with seven; or of taking careless landlords into court because they would not provide the proper garbage receptacles; or of arresting the tenant who tried to make the garbage wagons carry away the contents of his stable.

With the two or three residents who nobly stood by, we set up six of those doleful incinerators which are supposed to burn garbage with the fuel collected in the alley itself. The one factory in town which could utilize old tin cans was a window weight factory, and we deluged that with ten times as many tin cans as it could use—much less pay for.

Source: Jane Addams, *Twenty Years at Hull-House* (New York: Macmillan, 1911), pp. 281, 285–86.

---

## DOCUMENT 73: Gifford Pinchot on Conservation and the National Interest (1911)

Gifford Pinchot, the first professional American forester, was appointed head of the Department of Agriculture's Forest Service in 1894. In his early years with the service, he focused his attention on encouraging the "production of the largest amount of the most valuable timber in the shortest time on a given area,"[3] but as he became embroiled with government administrators who gave little thought to the consequences of resource exploitation and waste, he became a fierce conservation advocate.

After he instituted charges against President William Howard Taft's secretary of the Interior, Richard Ballinger [see Document 69], for reversing Theodore Roosevelt's conservation policies, Pinchot was dismissed from the service by Taft for insubordination. He later served under President Franklin D. Roosevelt as head of the U.S. Forestry Service.

The conservation of our natural resources is a question of primary importance on the economic side. It pays better to conserve our natural resources than to destroy them, and this is especially true when the national interest is considered. But the business reason, weighty and worthy though it be, is not the fundamental reason. In such matters, business is a poor master but a good servant. The law of self-preservation is higher than the law of business, and the duty of preserving the Nation is still higher than either.

The American Revolution had its origin in part in economic causes, and it produced economic results of tremendous reach and weight. The Civil War also arose in large part from economic conditions, and it has had the largest economic consequences. But in each case there was a higher and more compelling reason. So with the third great crisis of our history. It has an economic aspect of the largest and most permanent importance, and the motive for action along that line, once it is recognized, should be more than sufficient. But that is not all. In this case, too, there is a higher and more compelling reason. The question of the conservation of natural resources, or national resources, does not stop with being a question of profit. It is a vital question of profit, but what is still more vital, it is a question of national safety and patriotism also.

We have passed the inevitable stage of pioneer pillage of natural resources. The natural wealth we found upon this continent has made us rich. We have used it, as we had a right to do, but we have not stopped there. We have abused, and wasted, and exhausted it also, so that there is the gravest danger that our prosperity to-day will have been bought at the price of the suffering and poverty of our descendants.

*Source*: Gifford Pinchot, *The Fight for Conservation* (Garden City, NY: Doubleday, Page, 1911), pp. 126–28.

# Part V

---

# Rethinking Our Relationship to Nature, 1920–1960

The end of World War I brought with it a slackening of economic and industrial constraints and a surging demand for consumer goods. Within just a few years, a mass consumption economy emerged. By the 1930s, hundreds of thousands of people had acquired telephones, cars, phonographs, radios, and a host of other products that had come on the market in recent decades. During the Roaring Twenties a get-rich-quick mentality bred scandal and corruption. President Warren Harding's inability to rein in his underlings led to the Teapot Dome scandal of 1921, during which the valuable naval oil reserves in Teapot Dome, California, and Elk Hills, Wyoming, were transferred from the Navy Department to the Interior Department and then leased to two oil men, Harry Sinclair and Edward Doheny, after they paid a $100,000 bribe to Secretary of the Interior Albert Fall.

## URBAN GROWTH AND RURAL DEVELOPMENT

By the early 1920s, the U.S. population had reached 1 million, with more than half of the people living in urban areas. The new urban residents not only provided a huge market for America's expanding industry, but also pushed the borders of cities into what had once been farmland, making it necessary to transport agricultural produce longer distances, and placed increasing demands on metropolitan water supplies and waste disposal systems. As cars replaced horses on city streets, the manure problem disappeared. Eventually, however, it would be replaced by a new kind of pollution, auto emissions, but this would not become evident until after World War II. Although consumer buying slowed during the Great Depression, New Deal public works programs, such as the Tennessee Valley Authority (TVA) program, brought electrification to rural areas and in time swelled the demand for electrical appliances.

In the 1930s the dust bowl crisis in the south-central United States awakened America to the wastefulness of farming practices that resulted in devastating soil erosion. John Steinbeck in his novel *The Grapes of Wrath* [see Document 84] decried both the environmental and the human degradation resulting from these practices. In 1935 the Soil Conservation Service was established within the Department of the Interior to combat soil erosion and water wastage. The creation of the Civilian Conservation Corps in 1937 by President Franklin D. Roosevelt was both a response to the Great Depression—which began with the stock market crash of 1929 and was exacerbated by the dust bowl crisis—and an attempt to redress the country's prodigal use of its natural resources. Roosevelt had long recognized the relationship between resource protection and national well-being [see Document 77].

## THE SYNTHETIC ENVIRONMENT

Expansion of the chemical and food processing industries in the 1920s and 1930s brought a wide range of new products into existence to satisfy both the agricultural sector and the urban market. While many of these new, man-made substances provided great benefits for farmers and the general public, some of them had serious unanticipated negative consequences, including the endangerment of human health and the destruction of wildlife. New factories, unhampered by any kind of governmental environmental constraints, also made existing air and water pollution problems worse.

As the century progressed, opposition among the public to the government's leniency concerning industrial pollution and waste grew [see Document 78], and there was increasing demand for greater government oversight of the quality of manufactured goods and the labeling of processed foods and drugs [see Document 79]. The passage of the Food, Drug, and Cosmetics Act of 1938 strengthened the federal government's ability to clamp down on the free-wheeling sales of manufactured products to consumers.

In general, though, inadequate consideration was given to the long-term effects of the many new products that came to market in the 1930s and 1940s, such as pesticides like DDT and the numerous synthetic materials, including nylon and plastic, that were developed for use in World War II. It took many years before the impact of certain pesticides on the food chain and the problem of nonbiodegradable waste became evident.

## THE GROWTH OF SUBURBS

After World War II, home building boomed. Many of the new homes were constructed in the suburbs on former farmland, woodland, and open space, far from convenient shopping and transportation, and these suburbs spawned a new way of life based on the automobile. With the disappearance of farmland close to city centers, milk, poultry, and other fresh foods for the metropolitan areas had to be brought in from ever more distant farms. While some visionaries worried about the loss of open space and farmland in the 1950s—the Nature Conservancy was founded in 1951—and there were a few state and local efforts to protect farmland in the 1960s and 1970s [see Document 103], urban and suburban sprawl did not become a national issue until the early 1980s.

As the population rose, automobile use increased, highways spread across the nation, and suburban areas encroached on once-rural watersheds. In more and more places, the air was unfit to breathe and the water unsafe to drink. In 1955 the federal government began to finance and develop programs to prevent and control air pollution [see Document 93]. Effective federal water pollution control [see Document 115] and safe drinking water acts, though, were not passed until 1972 and 1974, respectively.

## REVAMPING THE HUMAN-NATURE RELATIONSHIP

By the 1920s, recognition of the decline in wildlife and wilderness areas had encouraged hunters, fishermen, and other sportsmen, including members of the Boone and Crockett Club, to reexamine their relationship with the wild. In 1925 George Bird Grinnell and Charles Sheldon noted, "The original purpose of the Boone and Crockett Club, to make hunting easier and more successful, has changed with changing conditions, so that now it is devoted chiefly to setting better standards in conservation."[1] The drop in numbers of certain wildlife species [see Document 86] and the disappearance of wilderness areas provided impetus for the creation of a host of new conservation organizations, including the Izaak Walton League (1922), the Wilderness Society (1935), Ducks Unlimited (1937), Defenders of Wildlife (1947), and the Conservation Foundation (1947), and by 1959 there was strong support for a Wilderness Act [see Document 95].

Shortly after the turn of the century, local governments, in an effort to contain building growth and industrial development, had begun to impose land use zoning regulations. By the 1920s, real estate, mining, and industrial interests started to feel the sting of these regulations and appealed to the courts for a redress of grievances [see Documents 74 and 75]. In general, the courts tended to side with the interests of big business, but over time, zoning laws have become increasingly stringent and have forced individual property owners as well as businesses of every type to adjust to communal interests. By the 1930s local and regional planning [see Document 83] had made great strides in bringing order to land development, but even in the 1950s insufficient thought was given to the consequences of building in areas with inadequate water supplies [see Document 92].

Much of the nation looked with favor on the transformation of the natural landscape, as industry expanded, dams were built, wetlands reclaimed, and forests and farmland paved over and turned into roads and housing developments. But a small segment of the population was horrified by the complete lack of regard for the needs of nature. In 1928 Henry Beston [Document 76] proposed that we rethink our relationship to other living things, and in 1933 Luther Standing Bear [Document 80] pointed out that respect for other living things is fundamental to the Native American view of life. Arthur Tansley [see Document 81] introduced the idea of the ecosystem in 1935, and the following year H. V. Harlan and M. L. Martini [see Document 82] raised concerns about decreasing biodiversity. In 1947, Marjory Stoneman Douglas [see Document 85], in an evocative book on the Everglades, called into question the policy of draining wetlands to reclaim land for agriculture and building development. By the time Fairfield Osborn proclaimed that the time had come for Americans to take note of the interrelatedness of all living things [see Document 87] and Aldo Leopold [see Document 88] proposed a new "land ethic"—calling for a change in the role of humans from that of conquerors to that of simple members of a community whose other members include water, soil, plants, and animals—the foundations for an environmental movement, expanded from the conservation movement of the turn of the century, had been laid.

## PEOPLE AND NATURE IN THE NUCLEAR AGE

One of the reactions to the devastation caused by the atomic bombs dropped on Hiroshima and Nagasaki in 1945 in the process of bringing World War II to a conclusion was a reconsideration of the uses of technology [see Document 91]. As people came to understand that they had the ability to destroy life on earth on an unprecedented scale, a litany of voices arose to plead for the preservation of the earth and the living things that inhabit it. Some of the voices belonged to anti-nuclear weapons activists, while others were those of individuals and groups protesting the waste and destruction of the nations' irreplaceable resources.

Forty years after Teddy Roosevelt had attempted to set up a commission to study the nation's resources and establish a sensible resource policy [see Document 70], President Harry Truman created the Materials Policy Commission with much the same objective. This time, adequate funding was made available. Although the commission's report commented on the dwindling supplies of many natural resources, it nevertheless recommended that the main goal of a U.S. materials policy should be the development of a sufficient supply of resources to ensure economic growth [see Document 89]. There was swift reaction to the report, with conservationists like Samuel Ordway calling for sustainable development [Document 90] and economists like John Kenneth Galbraith questioning the right of the United States to continue to consume resources at an inordinately high rate [Document 94].

## DOCUMENT 74: *Pennsylvania Coal Company v. Mahon et al.* (1922)

In 1878 the Pennsylvania Coal Company, the owner of land containing coal deposits, deeded the surface of some of its property with the express reservation that the company had the right to remove all the coal beneath the surface land. However, on May 27, 1921, the Pennsylvania legislature passed the Kohler Act, which prohibited the mining of anthracite coal within city limits in a way that would "cause the . . . subsidence of any dwelling or other structure used as a human habitation, or any factory, store, or other industrial or mercantile establishment in which human labor is employed," as well as the subsidence of any public street.

When the owners of a house constructed on land that had been deeded by the Pennsylvania Coal Company tried to prevent the coal company from mining under their building, citing the Kohler Act, the company sued the house owners. The case, which eventually reached the Supreme Court, confronted the issue of whether the government had a right to impose regulations that diminished the value of private property, and determined that "if regulation goes too far it will be recognized as a taking for which compensation must be paid." Justice Louis Brandeis, in his dissenting opinion, felt that regulation in the public interest that placed restrictions on the use of land in order to protect the public but that did not "appropriate or make any use of" the land is not a "taking." The ruling of the Court, however, indicated a bias in favor of large commercial interests as opposed to individual property holders.

## A. The Case and the Court's Ruling

This is a bill in equity brought by the defendants in error to prevent the Pennsylvania Coal Company from mining under their property in such way as to remove the supports and cause a subsidence of the surface and of their house. The bill sets out a deed executed by the Coal Company in 1878, under which the plaintiffs claim. The deed conveys the surface, but in express terms reserves the right to remove all the coal under the same, and the grantee takes the premises with the risk, and waives all claim for damages that may arise from mining out the coal. . . .

The statute [the Kohler Act] forbids the mining of anthracite coal in such way as to cause the subsidence of, among other things, any structure used as a human habitation, with certain exceptions, including among them land where the surface is owned by the owner of the underlying coal and is distant more than one hundred and fifty feet from any improved property belonging to any other person. As applied in this case the statute is admitted to destroy previously existing rights of property and contract. The question is whether the police power can be stretched so far.

Government hardly could go on if to some extent values incident to property could not be diminished without paying for every such change in the general law. As long recognized, some values are enjoyed under an implied limitation and must yield to the police power. But obviously the implied limitation must have its limits, or the contract and due process clauses are gone. One fact for consideration in determining such limits is the extent of the diminution. When it reaches a certain magnitude, in most if not in all cases there must be an exercise of eminent domain and compensation to sustain the act. So the question depends upon the particular facts. . . .

This is the case of a single private house. . . . A source of damage to such a house is not a public nuisance even if similar damage is inflicted on others in different places. The damage is not common or public.

. . .

The rights of the public in a street purchased or laid out by eminent domain are those that it has paid for. If in any case its representatives have been so short sighted as to acquire only surface rights without the right of support, we see no more authority for supplying the latter without compensation than there was for taking the right of way in the first place and refusing to pay for it because the public wanted it very much. The protection of private property in the Fifth Amendment presupposes that it is wanted for public use, but provides that it shall not be taken for such use without compensation.

## B. Justice Louis Brandeis's Dissenting Opinion

Coal in place is land; and the right of the owner to use his land is not absolute. He may not so use it as to create a public nuisance; and uses, once harmless, may, owing to changed conditions, seriously threaten the public welfare. Whenever they do, the legislature has power to prohibit such uses without paying compensation; and the power to prohibit extends alike to the manner, the character and the purpose of the use. . . .

Every restriction upon the use of property imposed in the exercise of the police power deprives the owner of some right theretofore enjoyed, and is, in that sense, an abridgment by the State of rights in property without making compensation. But restriction imposed to protect the public health, safety or morals from dangers threatened is not a taking. The restriction here in question is merely the prohibition of a noxious use. The property so restricted remains in the possession of its owner. The State does not appropriate it or make any use of it. The State merely prevents the owner from making a use which interferes with the paramount rights of the public. Whenever the use prohibited ceases to be noxious,—as it may because of further change in local or social conditions,—the restriction will have to be removed and the owner will again be free to enjoy his property as heretofore.

The restriction upon the use of this property can not, of course, be lawfully imposed, unless its purpose is to protect the public.

*Source: United States Reports*, Vol. 260 (Washington, D.C.: Government Printing Office, 1923), pp. 412, 413, 415, 417.

---

## DOCUMENT 75: *Village of Euclid et al. v. Ambler Realty Company* (1926)

The earliest zoning laws were instituted at the beginning of the twentieth century by local governments attempting to limit building construction and the use of land in ways that were detrimental to the interests of the local community. In recent years zoning laws have been used as an important means of containing growth near and preventing the development of wetlands and other environmentally sensitive or ecologically important areas. The *Euclid* (Ohio) case established that communities have the right to enact zoning laws in the public interest even if they cause the property of certain individuals to be devalued.

Appeal from a decree of the District Court enjoining the Village and its Building Inspector from enforcing a zoning ordinance. The suit was brought by an owner of unimproved land within the corporate limits of the village, who sought the relief upon the ground that, because of the building restrictions imposed, the ordinance operated to reduce the normal value of his property, and to deprive him of liberty and property without due process of law.

\* \* \*

A motion was made in the court below to dismiss the bill on the ground that, because complainant [appellee] had made no effort to obtain a building permit or apply to the zoning board of appeals for relief as it might have done under the terms of the ordinance, the suit was premature. The motion was properly overruled. The effect of the allegations of the bill is that the ordinance of its own force operates greatly to reduce the value of appellee's lands and destroy their marketability for industrial, commercial and residential uses; and the attack is directed, not against any specific provision or provisions, but against the ordinance as an entirety. Assuming the premises, the existence and maintenance of the ordinance, in effect, constitutes a present invasion of appellee's property rights and a threat to continue it. . . .

. . . The question is . . . : Is the ordinance invalid in that it violates the constitutional protection "to the right of property in the appellee by attempted regulations under the guise of the police power, which are unreasonable and confiscatory?"

Building zone laws are of modern origin. They began in this country about twenty-five years ago. Until recent years, urban life was comparatively simple; but with the great increase and concentration of population, problems have developed, and constantly are developing, which require, and will continue to require, additional restrictions in respect of the use and occupation of private lands in urban communities. Regulations, the wisdom, necessity and validity of which, as applied to existing conditions, are so apparent that they are now uniformly sustained, a century ago, or even half a century ago, probably would have been rejected as arbitrary and oppressive. . . .

The ordinance now under review, and all similar laws and regulations, must find their justification, in some aspect of the police power, asserted for the public welfare. The line which in this field separates the legitimate from the illegitimate assumption of power is not capable of precise delimitation. It varies with circumstances and conditions. A regulatory zoning ordinance, which would be clearly valid as applied to the great cities, might be clearly invalid as applied to rural communities. In solving doubts, the maxim *sic utere tuo ut alienum non laedas* [use what is yours in such a way as not to cause harm to others], which lies at the foundation of so much of the common law of nuisances, ordinarily will furnish a fairly helpful clew. And the law of nuisances, likewise, may be consulted, not for the purpose of controlling, but for the helpful aid of its analogies in the process of ascertaining the scope of, the power. Thus the question whether the power exists to forbid the erection of a building of a particular kind or for a particular use, like the question whether a particular thing is a nuisance, is to be determined, not by an abstract consideration of the building or of the thing considered apart, but by considering it in connection with the circumstances and the locality. . . .

There is no serious difference of opinion in respect of the validity of laws and regulations fixing the height of buildings within reasonable limits, the character of materials and methods of construction, and the adjoining area which must be left open, in order to minimize the danger of fire or collapse, the evils of over-crowding, and the like, and excluding from residential sections offensive trades, industries and structures likely to create nuisances. . . .

Here, however, the exclusion is in general terms of all industrial establishments, and it may thereby happen that not only offensive or dangerous industries will be excluded, but those which are neither offensive nor dangerous will share the same fate. . . .

It is said that the Village of Euclid is a mere suburb of the City of Cleveland; that the industrial development of that city has now reached and in some degree extended into the village and, in the obvious course of things, will soon absorb the entire area for industrial enterprises; that the effect of the ordinance is to divert this natural development elsewhere with the consequent loss of increased values to the owners of the lands within the village borders. But the village, though physically a suburb of Cleveland, is politically a separate municipality, with powers of its own and authority to govern itself as it sees fit within the limits of the organic law of its creation and the State and Federal Constitutions. Its governing authorities, presumably representing a majority of its inhabitants and voicing their will, have determined, not that industrial development shall cease at its boundaries, but that the course of such development shall proceed within definitely fixed lines. If it be a proper exercise of the police power to relegate industrial establishments to localities separated from residential sections, it is not easy to find a sufficient reason for denying the power because the effect of its exercise is to divert an industrial flow from the course which it would follow, to the injury of the residential public if left alone, to another course where such injury will be obviated. It is not meant by this, however, to exclude the possibility of cases where the general public interest would so far outweigh the interest of the municipality that the municipality would not be allowed to stand in the way.

[We have determined] that the ordinance in its general scope and features so far as its provisions are here involved, is a valid exercise of authority.

*Source: United States Reports*, Vol. 272 (Washington, D.C.: Government Printing Office, 1927), pp. 367, 386–90, 397.

---

## DOCUMENT 76: Henry Beston on the Human Relationship with Nature (1928)

Henry Beston, like Henry Thoreau and John Muir before him, sought isolation and nearness to nature. Having bought fifty acres and built a two-room cottage among the dunes of Nauset Beach on the eastern shore of Cape Cod, facing the open Atlantic, he planned to spend two weeks there in the fall of 1926, but captivated by "the beauty and mystery of this earth and outer sea,"[2] he stayed for a year and wrote *The Outermost House*. In the book, which tells about his year on the beach, Beston proposed that humans need to develop a new relationship with their environment. Rachel Carson [see Document 99] cited *The Outermost House* as the book that most profoundly influenced her writing, and federal officials noted the book's role in inspiring the creation of the Cape Cod National Seashore.

We need another and a wiser and perhaps a more mystical concept of animals. Remote from universal nature, and living by complicated artifice, man in civilization surveys the creatures through the glass of his knowledge and sees thereby a feather magnified and the whole image in distortion. We patronize them for their incompleteness, for their tragic fate of having taken form so far below ourselves. And therein we err, and greatly err. For the animal shall not be measured by man. In a world older and more complete than ours they move finished and complete, gifted with extensions of the senses we have lost or never attained, living by voices we shall never hear. They are not brethren, they are not underlings; they are other nations, caught with ourselves in the net of life and time, fellow prisoners of the splendour and travail of the earth.

*Source*: Henry Beston, *The Outermost House* (New York: Henry Holt, 1992), pp. 24–25.

## DOCUMENT 77: Franklin D. Roosevelt on the Conservation of America's Forests (1930)

Franklin Delano Roosevelt, Theodore Roosevelt's second cousin, grew up in a family that venerated the outdoors and was sensitive to concerns about conservation. In the New York State legislature FDR served as chairman of the Senate Committee on Forest, Fish, and Game, and as governor of the state he endorsed conservationist efforts. FDR's conviction that the government should take a leadership role in encouraging beneficial social change was probably responsible for his greatest contributions to the conservation movement, the creation of the Civilian Conservation Corps and the Soil Conservation Service.

Stones, steel, concrete and asphalt, bricks and glass meet the eye at every turn in the cities of America. Yet the products of the forest continue indispensable to the structure of our civilization.

A still heavier stake in intelligent conservation of this country's forest resources is held by the people of rural America. Agriculture uses more wood than any other industry. The people in the country benefit most from the prevention of floods. They own the lowlands where crops and property are destroyed. They own the highlands from which erosion sweeps away the fertility.

The small cities and towns, either as municipalities or through civic organizations, can make a notable contribution to progress in forest conservation. . . .

Back in 1911, as a greenhorn member of the New York State Senate, I happened to become chairman of the Committee on Forest, Fish, and Game. Anxious to stir up interest in my committee's work I arranged to have Gifford Pinchot [see Document 73] deliver a lecture in Albany. What he said I have probably forgotten but two pictures that he displayed are still vividly remembered.

He threw on the screen first the reproduction of a painting made in meticulous detail about the year 1400 by a Chinese artist. The scene was a beautiful valley in China, peopled with a city of a half million. Luxuriant crops in the carefully cultivated fields of the valley floor indicated a rich and well-tilled soil. A quiet river wound along, with indications on the bank that this was a stream of steady flow, free from periodic floods. A deep and dense forest of pine trees covered the mountains at either side of the valley. The whole scene was one of peace, prosperity and plenty.

Down the mountain at one side had been slashed a strip in which was a wooden trough, or flume, such as is used for sliding logs down a declivity. This was evidence that lumbering operations had been started.

Then Mr. Pinchot flashed on the screen a photograph of the same valley, made in 1900 from the identical spot occupied by the artist who five centuries before had painted the scene in photographic detail.

The mountain slopes had been completely denuded of their forests. Not a tree remained. Jutting rocks, deep gullies and barren spaces were there instead.

The whole valley floor was covered with a wilderness of rocks and bowlders that had been swept down by floods. No crops were growing, because no soil was left in which a seed might sprout. There was no river—only a dry stream bed where in season violent floods added to the destruction of what little was left to be damaged.

A poverty-stricken village of 5,000 remained within the still standing walls of the once prosperous city of a half million.

One need not be an alarmist to foresee that, without intelligent conservation measures, long before half a millennium passes some such contrasting pictures might be possible in our own United States. Even now we are consuming five times as much timber as is being grown. We plant in a year an area about equal to what is cut over in less than five days. Fortunately the federal government and many of the states are planning and working constructively to conserve the nation's timber resources.

A certain amount of sentiment clusters about trees and the forests and this I would not disparage, for I share it. We can, however, put that entirely aside, for the dollars and cents argument is powerful enough if we have the slightest consideration for future generations.

If no other fact were available to support this statement one would need only to point out that in eighty-five years lumber prices have increased three and a half times as rapidly as average prices of other common necessities. In consequence, in part at least, of this rise in price our per capita consumption of wood has declined by about forty per cent. Thirty-three of our states are now sending to other states for their lumber supplies.

We pay two hundred and fifty million dollars a year of freight because the remaining forests are so distant from the centers of consumption. The steel rails of our Northern transportation system are being laid on ties shipped from the far Northwest. Our newsprint supplies are coming from long distances. We use eight million tons of paper a year. It takes five million trees annually to support our telephone and telegraph wires. Wood is a staple necessity of everyday life.

Source: Franklin D. Roosevelt, "A Debt We Owe," Country Home 54 (June 1930): 12–13.

## DOCUMENT 78: Stuart Chase on Waste in the Machine Age (1931)

An author of popular books on economic problems in the United States, Stuart Chase urged that the country develop an industrial order in which production and distribution would be aimed at satisfying the needs of consumers—in the present and future—rather than of profit makers.

[One of the four] main channel[s] of waste is measured not in man-power, but in tonnage and horse-power. It is the measure of the gutting of the continent of North America. For every barrel of oil which has reached the pipe line, three barrels have been lost under ground; for every ton of coal which has come out of the pit, another ton has been left forever unreclaimable in the mine. We are cutting our forests four times faster than they are growing, which gives them—at this rate of exploitation—only another generation, while our effects at reforestation progress at a rate that will require 900 years to plant the land now idle and needing planting. Having cut the wood, we lose two-thirds of it in process of manufacture. Having made the paper, we use such wasteful sizes and grades that ninety-one business organizations taken at random have been found to throw $1,000,000 a year into the wastebasket.

In soils, fisheries, minerals, bird life, the neglect of waterpower—the story of ruthless pioneering, with no provision for the future, is repeated. . . . Meanwhile a single Sunday edition of the *New York Times*—75 per cent of which is advertising matter—consumes in wood pulp about 14 acres of forest land.

* * *

Modern industry, it is universally conceded, is operated on the basis of production for profit. The usefulness of the thing produced is a by-product. Realistic defenders of the present order admit this, but go on to explain that the profit motive provides so strong an incentive for pro-duction that more by way of consumable goods is thrown off—even as a by product—than could possibly be attained under any system founded on production for use only. In short, it is claimed that the way-faring man secures a greater net benefit from the profit system—despite its left-handed regard for his interests—than he could from any system designed directly to serve him.

*Source*: Stuart Chase, *Waste and the Machine Age* (New York: League for Industrial Democracy, 1931), pp. 34–36.

---

## DOCUMENT 79: Arthur Kallet and F. J. Schlink on the Dangers of Manufactured Products (1932)

---

> One of the early alarmist books about the dangers of commercial prod-
> ucts, Kallet and Schlink's *100,000,000 Guinea Pigs* created a public
> outcry that led to the passage of the Food, Drug, and Cosmetics Act of
> 1938 and resulted in major changes in food and drug policies.

In the magazines, in the newspapers, over the radio, a terrific verbal barrage has been laid down on a hundred million Americans, first, to set in motion a host of fears about their health, their stomachs, their bowels, their teeth, their throats, their looks; second, to persuade them that only by eating, drinking, gargling, brushing, or smearing with Smith's Whole Vitamin Breakfast Food, Jones' Yeast Cubes, Blue Giant Apples, Prussian Salts, Listroboris Mouthwash, Grandpa's Wonder Toothpaste, and a thousand and one other foods, drinks, gargles and pastes, can they either postpone the onset of disease, of social ostracism, of business failure, or recover from ailments, physical or social, already contracted.

If these foods and medicines were—to most of the people who use them—merely worthless; if there were no other charge to be made than that the manufacturers', sales managers', and advertising agents' claims for them were false, [*100,000,000 Guinea Pigs*] would not have been written. But many of them, including some of the most widely advertised and sold, are not only worthless, but are actually dangerous. That *All-Bran* you eat every morning—do you know that it may cause serious intestinal trouble? That big, juicy apple you have at lunch—do you know that indifferent Government officials let it come to your table coated with arsenic, one of the deadliest of poisons? The *Pebeco* Toothpaste with which you brush your teeth twice every day—do you know that a tube of it contains enough poison, if eaten, to kill three people; that, in fact, a German army officer committed suicide by eating a tubeful of this particular tooth paste? The *Bromo-Seltzer* that you take for headaches—do you know that it contains a poisonous drug which has been responsible for many deaths and, the American Medical Association says, at least one case of sexual impotence?

Using the feeble and ineffective pure food and drug laws as a smoke-screen, the food and drug industries have been systematically bombarding us with falsehoods about the purity, healthfulness, and safety of their products, while they have been making profits by experimenting on us with poisons, irritants, harmful chemical preservatives, and dangerous drugs.

... [W]e consumers are being forced into the role of laboratory guinea pigs through huge loopholes in obviously weak and ineffective laws.

*Source*: Arthur Kallet and F. J. Schlink, *100,000,000 Guinea Pigs: Dangers in Everyday Foods, Drugs, and Cosmetics* (New York: Vanguard Press, 1932), pp. 3–4.

## DOCUMENT 80: Luther Standing Bear on Native Americans and the Rights of Other Living Things (1933)

A Sioux chieftain and staunch advocate of Native American rights, Luther Standing Bear offers us a view of Native Americans as conservationists. Standing Bear's analysis of the Indians' perception of their relationship to other living things vividly depicts a mind-set at variance with the ideas of most other Americans.

The Indian was a natural conservationist. He destroyed nothing, great or small. Destruction was not a part of Indian thought and action; if it had been, and had the man been the ruthless savage he has been accredited with being, he would have long ago preceded the European in the labor of destroying the natural life of this continent. The Indian was frugal in the midst of plenty. When the buffalo roamed the plains in multitudes he slaughtered only what he could eat and these he used to the hair and bones. Early one spring the Lakotas were camped on the Missouri river when the ice was beginning to break up. One day a buffalo floated by and it was hauled ashore. The animal proved to have been freshly killed and in good condition, a welcome occurrence at the time since the meat supply was getting low. Soon another came floating downstream, and it was no more than ashore when other came into view. Everybody was busy saving meat and hides, but in a short while the buffalo were so thick on the water that they were allowed to float away. Just why so many buffalo had been drowned was never known, but I relate the instance as a boyhood memory.

I know of no species of plant, bird, or animal that were exterminated until the coming of the white man. For some years after the buffalo disappeared there still remained huge herds of antelope, but the hunter's work was no sooner done in the destruction of the buffalo than his attention was attracted toward the deer. They are plentiful now only where protected. The white man considered natural animal life just as he did the natural man life upon this continent, as "pests." Plants which the Indian found beneficial were also "pests." There is no word in the Lakota vocabulary with the English meaning of this word.

There was a great difference in the attitude taken by the Indian and the Caucasian toward nature, and this difference made of one a conservationist and of the other a non-conservationist of life. The Indian, as well as all other creatures that were given birth and grew, were sustained by the common mother—earth. He was therefore kin to all living things and he gave to all creatures equal rights with himself. Everything of earth was loved and reverenced.

* * *

From Wakan Tanka there came a great unifying life force that flowed in and through all things—the flowers of the plains, blowing winds, rocks, trees, birds, animals—and was the same force that had been breathed into the first man. Thus all things were kindred and brought together by the same Great Mystery.

Kinship with all creatures of the earth, sky, and water was a real and active principle. For the animal and bird world there existed a brotherly feeling that kept the Lakota safe among them. And so close did some of the Lakotas come to their feathered and furred friends that in true brotherhood they spoke a common tongue.

The animal had rights—the right of man's protection, the right to live, the right to multiply, the right to freedom, and the right to man's indebtedness—and in recognition of these rights the Lakota never enslaved the animal, and spared all life that was not needed for food and clothing.

*Source*: Luther Standing Bear, *Land of the Spotted Eagle* (Boston: Houghton Mifflin, 1933), pp. 165–66, 193.

## DOCUMENT 81: Arthur Tansley on the Concept of the Ecosystem (1935)

Arthur Tansley, a plant ecologist and one of the founders of the British Ecological Society (established in 1913), introduced the concept of the natural world as a set of complex, interacting communities. This selection is taken from an article by Tansley that appeared in the journal of the Ecological Society of America, which began publication in 1920, five years after the organization of the society.

[T]he more fundamental conception [than the plant biologist Frederic Clements' application of the term *biome* to "the whole complex of organisms inhabiting a given region"] is, as it seems to me, the whole *system* (in the sense of physics), including not only the organism-complex, but also the whole complex of physical factors forming what we call the environment of the biome—the habitat factors in the widest sense. Though the organisms may claim our primary interest, when we are trying to think fundamentally we cannot separate them from their special environment, with which they form one physical system.

It is the systems so formed which, from the point of view of the ecologist, are the basic units of nature on the face of the earth. Our natural human prejudices force us to consider the organisms (in the sense of the biologist) as the most important parts of these systems, but certainly the inorganic "factors" are also parts—there could be no systems without them, and there is constant interchange of the most various kinds within each system, not only between the organisms but between the organic and the inorganic. These *ecosystems*, as we may call them, are of the most various kinds and sizes. They form one category of the multitudinous physical systems of the universe, which range from the universe down to the atom.

*Source*: "The Use and Abuse of Vegetational Concepts and Terms," *Ecology* 16 (1935): pp. 295–99, in Carolyn Merchant, *Major Problems in American Environmental History* (Lexington, MA: Heath, 1992), p. 450.

---

## DOCUMENT 82: H. V. Harlan and M. L. Martini on the Loss of Genetic Diversity (1936)

H. V. Harlan, an agronomist, and M. L. Martini, a botanist, were working in the Department of Agriculture's Bureau of Plant Industry when they recognized the threat to genetic diversity posed by the widespread use of a few popular varieties of cereal grains. Although plant and animal breeding had been practiced on a small scale for thousands of years, it was not until the second half of the nineteenth century that it seriously began to affect the continued existence of species that had developed naturally.

Harlan and Martini were concerned about the loss of genetic resources and ancient plant varieties and about the effect on the development of new species. Within the next few decades, as scientists began to understand the relationship between biodiversity and ecosystem functioning, the wider ramifications of this loss of biodiversity became evident.

In the great laboratory of Asia, Europe, and Africa, unguided barley breeding has been going on for thousands of years. Types without num-

ber have arisen over an enormous area. The better ones have survived.
Many of the surviving types are old. Spikes from Egyptian ruins can
often be matched with ones still growing in the basins along the Nile.
The Egypt of the Pyramids, however, is probably recent in the history
of barley. . . . In the hinterlands of Asia there were probably barley fields
when man was young.

The progenies of these fields with all their surviving variations constitute the world's priceless reservoir of germ plasm. It has waited
through long centuries. Unfortunately, from the breeder's standpoint, it
is now being imperiled. Historically, the tribes of Asia have not been
overfriendly. Trade and commerce of a sort have always existed. They
have existed, however, on a scale so small that agriculture has been little
affected. Modern communication is a real threat. A hundred years ago,
when the grain crop of north Africa failed, the natives starved. Today,
in years of shortage, the French supply their dependent populations with
seed from California. Arab farmers in Mariout sometimes sell short to
European buyers and import seed grains from Palestine. In a similar way
changes are slowly taking place in more remote places. When new bar-
leys replace those grown by the farmers of Ethiopia or Tibet, the world
will have lost something irreplaceable.

*Source*: H. V. Harlan and M. L. Martini, "Problems and Results in Barley Breeding," in *U.S. Department of Agriculture Yearbook of Agriculture, 1936* (Washington,
D.C.: Government Printing Office, 1936), p. 317.

---

## DOCUMENT 83: Lewis Mumford on Regional Planning (1938)

The cultural historian Lewis Mumford advocated strict zoning regulations, regional planning, and communal land ownership as means to
prevent inappropriate land development. He believed that a primary
goal of the design of buildings, cities, communities, and public and
private space should be to tailor an appropriate fit between humankind
and the natural world. A generation later, Ian McHarg [see Document
109], a disciple of Mumford, developed the concept of environmental
impact.

Regional planning is essentially the effort to apply scientific knowledge and stable standards of judgment, justified by rational human values, to the exploitation of the earth. Such knowledge was deliberately
flouted in the opening up of the dry lands which have become the dust-
bowl of America, and the commonwealth has paid dearly for that sacrifice to the demands of the individual farmer and speculator. No

community can afford such luxuries of ignorance: the function of science is to reduce the area of such costly mistakes and finally wipe them out. Without the decisive control that rests with collective ownership, in the hands of responsible public administrators, working for the common good, regional planning is an all but impossible task: at best it must confine itself to weak admonitions, partial prohibitions, various forms of negative action: at most it can say what shall not be done, but it has little power to command the forces of positive action.

The common ownership of land would put the division and supervision of the land in the hands of the appropriate local and regional authorities, who would map out areas of cultivation, areas of mining, areas of urban settlement, as they now map out areas for public parks. On this basis, a stable social adjustment could be worked out for every part of the region, and for every type of resource and activity. This common ownership is not an objective in itself: it is merely a means toward creating a system of dressing and keeping the land as it must be dressed and kept for an advanced civilization. Something can indeed be done by education and public regulation where the obsolete system of private ownership and control is preserved; but infinitely more can and must be done by active authorities, capable not merely of suggestion but decisive action, capable of looking ahead over half a century or more, and borrowing funds on the basis of such long term action. A useful step toward this system of common ownership consists in a broadly applied scheme of land zoning: such as that provided in the current English law which "sterilizes" against change of use without public permission all existing rural areas, or like that worked out . . . in the Ruhr district before 1933. A partial step in the same direction—restricting marginal lands against inappropriate uses—has been made in Wisconsin.

*Source*: Lewis Mumford, *The Culture of Cities* (New York: Harcourt Brace, 1938), p. 329.

---

## DOCUMENT 84: John Steinbeck's *The Grapes of Wrath* (1939)

Like Jonathan Edwards, Thomas Jefferson, and the nineteenth-century transcendentalists, John Steinbeck believed that communion with the land was necessary for the well-being of the human spirit. He saw the dust bowl crisis of the 1930s, which wrought havoc on the lives of the small farmers of the prairie states, as a product of the rape of the land by agribusiness working hand in hand with the government.

His novel *The Grapes of Wrath* was a scathing attack on the destructive practices of agribusiness as well as on government land and water policies. In the book, Steinbeck paralleled the degradation of the lives of the Okies with the degradation of the prairie sod and depicted the tragic, systematic destruction of a fragile ecosytem.

The owner men sat in the cars and explained. You know the land is poor. You've scrabbled at it long enough, God knows.

The squatting tenant men nodded and wondered and drew figures in the dust, and yes, they knew, God knows. If the dust only wouldn't fly. If the top would only stay on the soil, it might not be so bad.

The owner men went on leading to their point: You know the land's getting poorer. You know what cotton does to the land; robs it, sucks all the blood out of it.

The squatters nodded—they knew, God knew. If they could only rotate the crops they might pump blood back into the land.

Well, it's too late. And the owner men explained the workings and the thinkings of the monster that was stronger than they were. A man can hold land if he can just eat and pay taxes; he can do that.

Yes, he can do that until his crops fail one day and he has to borrow money from the bank.

But—you see, a bank or a company can't do that, because those creatures don't breathe air, don't eat side-meat. They breathe profits; they eat the interest on money.

\* \* \*

The squatting men looked down again. What do you want us to do? We can't take less share of the crop—we're half starved now. The kids are hungry all the time. We got no clothes, torn an' ragged. If all the neighbors weren't the same, we'd be ashamed to go to meeting.

And at last the owner men came to the point. The tenant system won't work any more. One man on a tractor can take the place of twelve or fourteen families. Pay him a wage and take all the crop. We have to do it. We don't like to do it. But the monster's sick. Something's happened to the monster.

But you'll kill the land with cotton.

We know. We've got to take cotton quick before the land dies. Then we'll sell the land. Lots of families in the East would like to own a piece of land.

The tenant men looked up alarmed. But what'll happen to us? How'll we eat?

You'll have to get off the land. The plows'll go through the dooryard.

*Source*: John Steinbeck, *The Grapes of Wrath* (New York: Viking/Penguin, 1939), pp. 43, 44.

## DOCUMENT 85: Marjory Stoneman Douglas on the Everglades (1947)

The Florida Everglades, so eloquently described by the writer Marjory Stoneman Douglas as a "river of grass," is a unique area. This 1,700-square-mile wetland is part of a 9,000-square-mile ecosystem that includes the Kissimmee River basin and Lake Okeechobee and provides the biologic and hydrologic foundation for south Florida's economic prosperity.

Douglas posited that the draining, diking, channeling, and manipulation of the waters of this system to make way for agricultural lands, homes, and roads—which had been official state policy since the late 1800s—was a recipe for destroying the system. Although she felt that the newly designated Everglades National Park would inhibit further changes to the southern part of the system, she feared the power of special interests. Her warnings were largely ignored until the 1980s [see Document 131].

There are no other Everglades in the world.

They are, they have always been, one of the unique regions of the earth, remote, never wholly known. Nothing anywhere else is like them: their vast glittering openness, wider than the enormous visible round of the horizon, the racing free saltness and sweetness of their massive winds, under the dazzling blue heights of space. They are unique also in the simplicity, the diversity, the related harmony of the forms of life they enclose. The miracle of the light pours over the green and brown expanse of saw grass and of water, shining and slow-moving below, the grass and water that is the meaning and the central fact of the Everglades of Florida. It is a river of grass.

\* \* \*

[H]istory, the recorded time of the earth and of man, is in itself something like a river. To try to present it whole is to find oneself lost in the sense of continuing change. The source can be only the beginning in time and space, and the end is the future and the unknown. . . .

So it is with the Everglades, which have that quality of long existence in their own nature. They were changeless. They are changed.

They were complete before man came to them, and for centuries afterward, when he was only one of those forms which shared, in a finely balanced harmony, the forces and the ancient nature of the place.

\* \* \*

The most important single recommendation of the Everglades Project Reports [issued in 1946] was for a single plan of development and water control for the whole area, under the direction of a single engineer and his board. Only in that way could the conflicting demands of local areas be equalized, so that the soil fit for high cultivation could be used and maintained without detriment to the water supply of the lower areas. It would maintain areas for water conservation. It would control salt intrusion. A well-planned system of canals that would discharge excess lake water into the open Glades would permit the river of grass to flow again with sweet water.

The report was ... studied by the thoughtful men in the growing cities. Its recommendations received wide attention and support, although there was opposition by local interests with political power. The growers about the lake were still afraid of floods, and they felt that water control would endanger the lands already under production. The cattlemen of the west were still resentful of an over-all control. Many owners of great areas of mucklands, draining and diking and pumping their irrigation ditches full from the seaward-flowing canals, refused to consider anything more important than their own immediate profits and would fight more fiercely than any others a co-ordinated control.

It was too soon to expect that all these people would see that the destruction of the Everglades was the destruction of all.

*Source*: Marjory Stoneman Douglas, *The Everglades: River of Grass* (New York: Rinehart, 1947), pp. 5–6, 8–9, 383.

## DOCUMENT 86: Roger Tory Peterson on Bird Population (1948)

The noted ornithologist and author of the popular *Field Guides* Roger Tory Peterson raised people's awareness of how human activity, including the changes that people make in their environment, affects wildlife populations, forcing some species to dwindle and enabling others to multiply. His books offered an innovative means of identifying birds—one that eliminated the need to shoot and collect them in order to be certain of their identity. His books also moved interest in bird-watching—or "birding," as it is popularly referred to—from the pastime of a few hundred to the passion of 55 million Americans. For many people, birding proved to be but the first step in the development of environmental consciousness.

Every ornithologist I know would give his soul to step back into time and walk the [North American] continent in the historic year of 1492. It was all virgin country then, with trees centuries old and the native grass waist high on the prairies. The broad distributional concepts—the life zones or the biomes—probably would have been much more satisfactory in those days when most environments were in relatively stable "climaxes," as the old mature plant associations are called. But that was before man set in motion the constant chain of changes that take place wherever he goes. For civilized man is the great disturber. Some would call him a destroyer, but that I think is a harsh term. Certainly he brings change.

* * *

Let us look briefly at the score sheet and see which way our birds are going.

Most of the waterfowl and the native upland game birds are below par. There has been some restoration in places, but there were many more of them in primitive America. Even twentieth-century America could support more than it does. Hunting often exacts a greater toll than the traffic will bear.

Today all marsh birds, not waterfowl alone, are in the red. Every day plans are made somewhere in this country to drain a lake, a swamp or a marsh. The records show that close to 100,000,000 acres of land have been drained in the United States for agriculture alone. Other millions of acres have been ditched to control mosquitoes. Considering that a marsh or swamp habitat harbors nine or ten nesting birds per acre and most farming country an average of fewer than three, this means that at least half a billion birds may have been eliminated from the face of the continent by the simple process of digging ditches.

The birds of prey—the hawks and the owls—are much reduced. . . . [T]he gunner often blamed the growing scarcity of game not on himself but on the natural predators, which had lived in satisfactory adjustment to their prey for thousands of years. He called all hawks "vermin," competitors to be shot and destroyed. And the skies over most of America today are still not empty enough of hawks to satisfy him.

The vultures, on the other hand, seem to be spreading.

*Source*: Roger Tory Peterson, *Birds over America* (New York: Dodd, Mead, 1948), pp. 71–73.

## DOCUMENT 87: Fairfield Osborn on the Relatedness of All Living Things (1948)

Fairfield Osborn, who founded the Conservation Foundation in 1947 to raise public consciousness about ecological problems, contended that the idea that humans could replace the fundamental functioning of natural systems was a dangerous illusion and warned of a "silent spring" more than a dozen years before Rachel Carson published her popular book *Silent Spring* [see Document 99].

There is no risk in making the flat statement that in a world devoid of other living creatures, man himself would die. This fact—call it a theory if you will—is far more provable than the accepted theory of relativity. Involved in it is, in truth, another kind of principle of relativity—the relatedness of all living things.

As a somewhat extreme illustration, among many others, take that form of life that man likes the least—of which the unthinking person would at once say, "Kill them all." Insects. Of the extraordinary number of kinds of insects on the earth—about three quarters of a million different species have already been identified—a small minority are harmful to man, such as the anopheles mosquito, lice, the tsetse fly, and crop-destroying insects. On the other hand, innumerable kinds are beneficent and useful. Fruit trees and many crops are dependent upon insect life for pollination or fertilization; soils are cultured and gain their productive qualities largely because of insect life. Human subsistence would, in fact, be imperiled were there no insects. On the other hand, insects, capable of incredibly rapid reproduction, have been freed by man himself of many natural controls such as those once provided by birds, now so diminished in numbers, or by fish, once a potent factor in insect control, no longer existing in countless lakes, rivers and streams now so polluted that aquatic life has disappeared. In attempting to find substitutes for natural controls man has resorted to the use of chemicals of increasing power. A few years ago arsenicals came into style—widely used in freeing fruit orchards of pests. So promising this method has seemed—but insidiously it sometimes results in destroying insect-eating birds; and after several seasons the ground itself, in many orchards, has become so impregnated with the poison that the trees are affected and their fruit-bearing capacity dwindles. More recently a powerful chemical known as D.D.T. seems the cure-all. Some of the initial experiments with this insect killer have been withering to bird life as a result of birds eating the insects that have been impregnated with the chemical. The careless use of D.D.T. can also result in destroying fishes, frogs and toads, all of which live on insects. This new chemical is deadly to many kind of insects—no doubt of that. But what of the ultimate and net result to the life scheme of the earth? On another front man is blindly in conflict with nature, too often overlooking the fact that the animal life of the earth, its interrelationships, its preservation, are wrapped up directly with his own well-being. Will the day come when this is generally realized?

*Source*: Fairfield Osborn, *Our Plundered Planet* (Boston: Little, Brown, 1948), pp. 60–63.

---

## DOCUMENT 88: Aldo Leopold's Land Ethic (1949)

Aldo Leopold began his professional career as a U.S. forester in the Southwest, and during these years he helped develop a national forest policy that included setting aside some forested lands as permanently protected areas. After moving to Wisconsin, he became involved with wildlife management, and initially, accepting the conventional wisdom of the period, Leopold considered predator control to be the primary object of his work. Eventually, though, he came to see good game management as a much more complex problem and became a champion of the concept of ecology. In *A Sand County Almanac*, from which the following selection is taken, Leopold provides us not only with a better understanding of the interrelatedness of all living things but also with a new set of rules, a new ethic, to govern human behavior toward the environment.

The first ethics dealt with the relation between individuals; the Mosaic Decalogue is an example. Later accretions dealt with the relation between the individual and society. The Golden Rule tries to integrate the individual to society; democracy to integrate social organization to the individual.

There is as yet no ethic dealing with man's relation to land and to the animals and plants which grow upon it. Land, like Odysseus' slave-girls, is still property. The land-relation is still strictly economic, entailing privileges but not obligations.

The extension of ethics to this third element in human environment is, if I read the evidence correctly, an evolutionary possibility and an ecological necessity. It is the third step in a sequence. The first two have already been taken. Individual thinkers since the days of Ezekiel and Isaiah have asserted that the despoliation of land is not only inexpedient but wrong. Society, however, has not yet affirmed their belief. I regard the present conservation movement as the embryo of such an affirmation.

An ethic may be regarded as a mode of guidance for meeting ecological situations so new or intricate, or involving such deferred reactions, that the path of social expediency is not discernible to the average individual. Animal instincts are modes of guidance for the individual in meeting such situations. Ethics are possibly a kind of community instinct in-the-making.

All ethics so far evolved rest upon a single premise: that the individual is a member of a community of interdependent parts. His instincts prompt him to compete for his place in that community, but his ethics prompt him also to co-operate (perhaps in order that there may be a place to compete for).

The land ethic simply enlarges the boundaries of the community to include soils, waters, plants, and animals, or collectively: the land.

This sounds simple: do we not already sing our love for and obligation to the land of the free and the home of the brave? Yes, but just what and whom do we love? Certainly not the soil, which we are sending helter-skelter downriver. Certainly not the waters, which we assume have no function except to turn turbines, float barges, and carry off sewage. Certainly not the plants, of which we exterminate whole communities without batting an eye. Certainly not the animals, of which we have already extirpated many of the largest and most beautiful species. A land ethic of course cannot prevent the alteration, management, and use of these "resources," but it does affirm their right to continued existence, and, at least in spots, their continued existence in a natural state.

In short, a land ethic changes the role of *Homo sapiens* from conqueror of the land-community to plain member and citizen of it. It implies respect for his fellow-members, and also respect for the community as such.

*Source*: Aldo Leopold, *Sand County Almanac* (New York: Oxford University Press, 1949), pp. 201–4.

## DOCUMENT 89: President's Materials Policy Commission on Economic Growth and Resource Policy (1952)

In recommending economic growth as the highest priority in the making of U.S. resource policy, despite the recognition of limitations on the availability of resources, the President's Materials Policy Commission revealed an unwillingness to upset the status quo. The report provoked the indignation of both conservationists and economists [see

Documents 90 and 94]. One of the outgrowths of this consternation was the formation in 1955 of Resources for the Future, a nonprofit corporation that studies resource scarcity.[3]

A hundred years ago, resources seemed limitless and the struggle upward from meager conditions of life was the struggle to create the means and methods of getting these materials into use. In this struggle we have succeeded so well that today, in thinking of expansion programs, full employment, new plants, or the design of a radical new turbine blade, too many of us blankly forget to look back to the mine, the land, the forest: the sources upon which we absolutely depend. So well have we built our high-output factories, so efficiently have we opened the lines of distribution to our remotest consumers that our sources are weakening under the constantly increasing strain of demand. As a Nation, we have always been more interested in sawmills than seedlings. We have put much more engineering thought into the layout of factories to cut up metals than into mining processes to produce them. We think about materials resources last, not first.

\* \* \*

The actions we as a Nation take or fail to take in meeting the materials problems in the period immediately ahead will affect profoundly the state of affairs many years hence. Upon our own generation lies the responsibility for passing on to the next generation the prospects of continued well-being.

\* \* \*

[The members of the President's Materials Policy Commission] share the belief of the American people in the principle of Growth. Granting that we cannot find any absolute reason for this belief we admit that to our Western minds it seems preferable to any opposite, which to us implies stagnation and decay. Where there may be any unbreakable upper limits to the continuing growth of our economy we do not pretend to know, but it must be part of our task to examine such apparent limits as present themselves.

... [W]e believe in private enterprise as the most efficacious way of performing industrial tasks in the United States. With this belief, a belief in the spur of the profit motive and what is called "the price system" obviously goes hand in hand.

\* \* \*

*The over-all objective of a national materials policy for the United States should be to insure an adequate and dependable flow of materials at the lowest cost consistent with national security and with the welfare of friendly nations.*

\* \* \*

As a Nation we have long lived and prospered mightily without serious concern for our material resources. Our sensational progress in production and consumption has been attributable not only to the freedom of our institutions and the enterprise of our people, but also to our spendthrift use of our rich heritage of natural resources. We have become the supreme advocates of the idea that man and his labor are the most valuable of all, and that inanimate materials are to be used as fully as possible to give men the greatest amount of return for the effort they put forth.

This still is and should be our goal, but the time has clearly passed when we can afford the luxury of viewing our resources as unlimited and hence taking them for granted. In the United States the supplies of the evident, the cheap, and the accessible (chemically and geologically) are running out. The plain fact seems to be that we have skimmed the cream of our resources as we now understand them; there must not be, at this decisive point in history, too long a pause before our understanding catches up with our needs. . . .

Growth of demand is at the core of the materials problem we face; it is the probability of continued growth, even more than the incursions of past growth and two world wars, that present us now with our long-range problems. It is mainly our unwillingness to stand still, to accept the status of a "mature economy," that challenges the adequacy of our resources.

*Source*: President's Materials Policy Commission, *Resources for Freedom*, Vol. 1: *Foundations for Growth and Security* (Washington, D.C.: President's Materials Policy Commission, June 1952), pp. 1, 2, 4, 5.

---

## DOCUMENT 90: Samuel H. Ordway, Jr., on Limits to Growth (1953)

Samuel Ordway, a lawyer and proactive member of the boards of several conservation organizations, including the Conservation Foundation, the Natural Resources Council of America, and the American Conservation Association, was one of the first environmental analysts to discuss limits to industrial growth. In the 1970s, Ordway's ideas on growth limitation were expanded on by a number of economists such as Donella Meadows, who asserted that, if civilization is to survive, limits must be placed on industrial investment as well as population growth.[4] By the 1980s the concept of limits to industrial growth had become part of the basic philosophy of such radical environmentalists as Mark Sagoff [see Document 124] and Arne Naess [see Document 128].

The end of expansion, if unexpected and involuntary, would mean the reversal of a major facet of our faith; it would impose a forced revision of ideology; it would mean mass discouragement and unemployment—it could mean revolution and dictatorship. It would not mean a return to pioneering—for there is no new land to pioneer. It would not mean a return from the city to the farm. There are too many people, trained in too many white and blue collar skills, to support themselves in the future by farming individually our remaining productive land. It might even force displaced segments of the American people to set to work clearing needed land of unused factories, roads, and airports and concentrate all our resources on restoring that land to productivity. It is difficult to believe that any such adjustment would be possible without conscription, work relief, and substantial loss of freedom and independence.

The industrial pattern of today has brought us health, wealth and leisure in unprecedented degree. It has made us reliant on machines and government—less reliant on ourselves. Success has led us to believe that the earth is a cornucopia, and the machine a god. It has led us to a false faith in man's omnipotence.

Malthus [see Document 26], were he living today, would have to modify his contention that it is [the] increasing number of mouths to feed that will exhaust the earth's food supplies and cause ultimate misery. He would probably say now, in the light of experience, that if the total of popular and industrial consumption of the produce of the earth exceeds reproduction over a long enough period, misery will be inevitable.

Our problem today is to be sure that in the long run we limit consumption to continuing supply. This can be done on a voluntary basis if we begin soon enough to distinguish needs from wants. Indeed, consumption can well continue to increase as the margin of supply increases. It is fundamental, however, that the continued betterment of man's level of living will depend on adjusting supply and demand. That is the new concept of conservation. Consumption cannot run beyond supply for long.

How to limit consumption to supply and maintain free enterprise is a matter of land-use planning, distribution of raw materials and rationing by industry, government and technology cooperating.

*Source*: Samuel H. Ordway, Jr., *Resources and the American Dream: Including a Theory of the Limit of Growth* (New York: Ronald Press, 1953), pp. 46–48.

## DOCUMENT 91: J. Robert Oppenheimer on the Use of Science (1953)

J. Robert Oppenheimer, one of the early advocates of the international control of atomic energy, led the Manhattan Project that developed the atomic bomb in 1945. In the late 1940s, however, when President Harry Truman wanted to develop a hydrogen bomb, Oppenheimer objected, and as a result of this opposition, Oppenhimer lost his security clearance during the McCarthy era.

Three hundred years after Francis Bacon had looked to science and technology to solve humanity's problems [see Document 8], Oppenheimer recognized that technology could be a force for evil as well as for good. This change in attitude toward science and technology reflects a growing awareness that humans cannot control nature and that some of our attempts to harness the forces of nature may prove disastrous in the long run.

We are today anxiously aware that the power to change is not always necessarily good.

As new instruments of war, of newly massive terror, add to the ferocity and totality of warfare, we understand that it is a special mark and problem of our age that man's ever-present preoccupation with improving his lot, with alleviating hunger and poverty and exploitation, must be brought into harmony with the over-riding need to limit and largely to eliminate resort to organized violence between nation and nation. The increasingly expert destruction of man's spirit by the power of police, more wicked if not more awful than the ravages of nature's own hand, is another such power, good only if never to be used.

We regard it as proper and just that the patronage of science by society is in large measure based on the increased power which knowledge gives. If we are anxious that the power so given and so obtained be used with wisdom and with love of humanity, that is an anxiety we share with almost everyone. But we also know how little of the deep new knowledge which has altered the face of the world, which has changed— and increasingly and ever more profoundly must change—man's views of the world, resulted from a quest for practical ends or an interest in exercising the power that knowledge gives. For most of us, in most of those moments when we were most free of corruption, it has been the beauty of the world of nature and the strange and compelling harmony of its order, that has sustained, inspirited, and led us.

*Source*: J. Robert Oppenheimer, "The Sciences and Man's Community," in *Science and the Common Understanding* (New York: Simon and Schuster, 1954), pp. 97–98.

## DOCUMENT 92: Bernard Frank on Development and Water Availability (1955)

Americans have often viewed water—whether for drinking, for industry, or for farm use—as an unlimited resource owed to us, almost as a constitutional right. Bernard Frank, who worked in the Watershed Management Research Division of the Forest Service, warns that we must give greater attention to water resource limitations.

The impact of new inventions and new developments and growth in population and industry [on water consumption] has not commonly been given the attention it has merited.

Many critical local water shortages therefore occurred that could have been forestalled. For example, rural electrification has brought about such heavy increases in the use of water for household and production purposes that the limited well-water supplies of many farms have been severely strained.

Similarly, factories have been built without prior studies to determine whether water would be available to operate the factories and to provide for the communities around them.

Towns, cities, industries, and farms have kept expanding beyond the safe limits of available water. Often makeshift efforts have been necessary to meet emergencies, especially in years of low rainfall. Such efforts have often hastened the depletion of the limited reserves in underground reservoirs, generated disputes with other cities or industries drawing on the same sources of water, introduced conflicts with the use of water for recreation, and threatened the permanent flooding of lands valuable for farming, forestry, wilderness, or wildlife.

. . . Use continues to rise; advancing standards of health and comfort, the application of more intensive farming practices, and the development of new products all impose additional demands.

*Source*: Bernard Frank, "Our Need for Water," in *Water: Agricultural Yearbook of 1955* (Washington, D.C.: U.S. Department of Agriculture, 1955), pp. 4–5.

## DOCUMENT 93: Clean Air Act (1955)

Air pollution and air quality have been a problem in the Western world since the beginning of the Industrial Revolution, and Americans have been concerned about these issues since the early nineteenth century [see Documents 41 and 67]. Until 1955, however, responsibility for clean air in the United States rested with states and local communities, which frequently ignored the problem. In 1955, the federal government—having recognized that the smoke produced in the Midwest moved with the air as it flowed from west to east, ultimately causing pollution problems in the East—took its first, tentative steps to encourage a nationwide advance toward clean air by funding state and local air pollution prevention and control programs.

### A. Congressional Findings and Declarations

The Congress finds—

(1) that the predominant part of the Nation's population is located in its rapidly expanding metropolitan and other urban areas, which generally cross the boundary lines of local jurisdictions and often extend into two or more States;

(2) that the growth in the amount and complexity of air pollution brought about by urbanization, industrial development, and the increasing use of motor vehicles, has resulted in mounting dangers to the public health and welfare, including injury to agricultural crops and livestock, damage to and the deterioration of property, and hazards to air and ground transportation;

(3) that air pollution prevention (that is, the reduction or elimination, through any measures, of the amount of pollutants produced or created at the source) and air pollution control at its source is the primary responsibility of States and local governments; and

(4) that Federal financial assistance and leadership is essential for the development of cooperative Federal, State regional, and local programs to prevent and control air pollution.

The purposes of this Act are—

(1) to protect and enhance the quality of the Nation's air resources so as to promote the public health and welfare and the productive capacity of its population;

(2) to initiate and accelerate a national research and development program to achieve the prevention and control of air pollution;

(3) to provide technical and financial assistance to State and local

governments in connection with the development and execution of their air pollution prevention and control programs; and

(4) to encourage and assist the development and operation of regional air pollution prevention and control programs.

## B. The Act

*Be it enacted* ... That ... it is hereby declared to be the policy of Congress to preserve and protect the primary responsibilities and rights of the States and local governments in controlling air pollution, to support and aid technical research to devise and develop methods of abating such pollution, and to provide Federal technical services and financial aid to State and local government air pollution control agencies and other public or private agencies and institutions in the formulation and execution of their air pollution abatement research programs. To this end, the Secretary of Health, Education, and Welfare and the Surgeon General of the Public Health Service (under the supervision and direction of the Secretary of Health, Education, and Welfare) shall have the authority relating to air pollution control vested in them respectively by this Act.

SEC. 2. (a) The Surgeon General is authorized, after careful investigation and in cooperation with other Federal agencies, with State and local government air pollution control agencies, with other public and private agencies and institutions, and with the industries involved, to prepare or recommend research programs for devising and developing methods for eliminating or reducing air pollution. For the purpose of this subsection the Surgeon General is authorized to make joint investigations with any such agencies or institutions.

(b) The Surgeon General may (1) encourage cooperative activities by State and local governments for the prevention and abatement of air pollution; (2) collect and disseminate information relating to air pollution and the prevention and abatement thereof; (3) conduct in the Public Health Service, and support and aid the conduct by State and local government air pollution control agencies, and other public and private agencies and institutions of, technical research to devise and develop methods of preventing and abating air pollution; and (4) make available to State and local government air pollution control agencies, other public and private agencies and institutions, and industries, the results of surveys, studies, investigations, research, and experiments relating to air pollution and the prevention and abatement thereof.

SEC. 3. The Surgeon General may, upon request of any State or local government air pollution control agency, conduct investigations and research and make surveys concerning any specific problem of air pollution confronting such State or local government air pollution control agency with a view to recommending a solution of such problem.

Sec. 4. The Surgeon General shall prepare and publish from time to time reports of such surveys, studies, investigations, research, and experiments. . . .

Sec. 5. (a) There is hereby authorized to be appropriated to the Department of Health, Education, and Welfare for each of the five fiscal years during the period beginning July 1, 1955, and ending June 30, 1960, not to exceed $5,000,000 to enable it to carry out its functions under this Act and, in furtherance of the policy declared in the first section of this Act, to (1) make grants-in-aide to State and local government air pollution control agencies, and other public and private agencies and institutions, and to individuals, for research, training, and demonstration projects, and (2) enter into contracts with public and private agencies and institutions and individuals for research, training, and demonstration projects.

*Source*: **A.** *United States Annotated Code*, Title 42: *The Public Health and Welfare* (St. Paul, MN: West Publishing, 1995), SS7401, July 14, 1955, pp. 27–28. **B.** Public Law 159, *United States Statutes at Large*, Vol. 69 (Washington, D.C.: Government Printing Office, 1955), 84th Cong., 1st sess., chap. 360, July 14, 1955, pp. 322–23.

---

## DOCUMENT 94: John Kenneth Galbraith Asks, "How Much Should a Country Consume?" (1958)

A professor of economics at Harvard who worked in the State Department Office of Economic Security Policy in the 1940s and served as ambassador to India during the Kennedy administration, John Kenneth Galbraith here lays out the dilemma confronting the United States in its attempt to be both a consumer society and a sustainable society.

What should be our policy toward consumption?

First, of course, we should begin to talk about it—and in the context of all its implications. It is silly for grown men to concern themselves mightily with supplying an appetite and close their eyes to the obvious and obtrusive question of whether the appetite is excessive.

If the appetite presents no problems—if resource discovery and the technology of use and substitution promise automatically to remain abreast of consumption and at moderate cost—then we need press matters no further. At least on conservation grounds there is no need to curb our appetite.

But to say this, and assuming that it applies comprehensively to both renewable and nonrenewable resources, is to say that there is no materials problem. It is to say that, except for some activities that by definition are noncritical, the conservationists are not much needed.

But if conservation is an issue, then we have no honest and logical course but to measure the means for restraining use against the means for insuring a continuing sufficiency of supply and taking the appropriate action. There is no justification for ruling consumption levels out of the calculation.

What would be the practical consequences of this calculation—taken honestly and without the frequent contemporary preoccupation not with solution but with plausible escape—I do not pretend to say.... I am impressed by the opportunities for resource substitution and by the contribution of technology in facilitating it. But the problem here is less one of theory than of technical calculation and projection.

*  *  *

[I]t would seem to me that any concern for materials use should be general. It should have as its aim the shifting of consumption patterns from those which have a high materials requirement to those which have a much lower requirement. The opportunities are considerable. Education, health services, sanitary services, good parks and playgrounds, orchestras, effective local government, a clean countryside, all have rather small materials requirements. I have elsewhere argued [in *The Affluent Society*] that the present tendency of our economy is to discriminate sharply against such production. A variety of forces, among them the massed pressures of modern merchandising, have forced an inordinate concentration of our consumption of what may loosely be termed consumer hardware. This distortion has been underwritten by economic attitudes which have made but slight accommodation to the transition of our world from one of privation to one of opulence. A rationalization of our present consumption patterns—a rationalization which would more accurately reflect free and unmanaged consumer choice—might also be an important step in materials conservation.

*Source*: John Kenneth Galbraith, "How Much Should a Country Consume?" in Henry Jarrett, ed., *Perspectives on Conservation: Essays on America's Natural Resources* (Baltimore: Johns Hopkins Press, 1958), pp. 98–99.

## DOCUMENT 95: David Brower Demands Support for the Wilderness Act (1959)

David Brower, part of a group that reinvigorated the Sierra Club and moved it to the forefront of environmental activism, raised our sensitivity to the importance of wilderness for the health of our civilization. The Wilderness Act was a great victory for the preservationist school of environmentalists, but by the time it was passed in 1964, the original text of the act had been changed in order to delay prohibitions on mining in wilderness areas. In the final version, the proposed wilderness council, which was to publish public reports on the status of the wilderness system, was also eliminated.

After leaving the Sierra Club in 1969, Brower founded Friends of the Earth and the Earth Island Institute.

The most important source of the vital organic forms constituting the chain of life is the gene bank that exists in wilderness, where the life force has gone on since the beginning uninterrupted by man and his technology. For this reason alone, it is important that the remnants of wilderness which we still have on our public lands be preserved by the best methods our form of government can find. The proposed National Wilderness Preservation System (now before the Congress) provides an excellent route to that goal, and especially dynamic leadership in the Congress and the administration will be required during the next decade to achieve the goal of wilderness preservation which the system would make possible. There will be important subsidiary benefits to recreation, to watershed protection, and to the beauty of the land.

A growing economy will have availed us nothing if it extinguishes our all-important wilderness. A gross misunderstanding of wilderness, in which it is evaluated according to the number of hikers who get into it, has been fostered for the past several years, to the great detriment of all the future. There must be no more needless, careless losses. There is no substitute for wilderness. What we now have is all that we shall ever have.

*Source*: David Brower, "The Meaning of Wilderness to Science," in Brower, *For Earth's Sake* (Salt Lake City: Peregrine Smith, 1990), pp. 271–72.

# Part VI

# The Heyday of the Environmental Movement, 1960–1980

The generation that came of age in the 1960s was the first generation of Americans to grow up watching television and escaping from the summer heat in air conditioned stores, offices, theaters, houses, or cars. It was a generation imbued with a sense of idealism, sparked by the short-lived Kennedy administration and Lyndon Johnson's dream of creating a "Great Society" without poverty or hunger. It was a generation inspired by the endless possibilities made visible by the dawn of the space age. It was also the first generation whose members had lived all their lives with the possibility of nuclear war looming over them like a dark cloud.

The sixties generation, which was predominantly urban and suburban (the 1960 U.S. population of 180 million was 70 percent urban) and the product of an era of affluence and consumerism, questioned not only its parents' wholehearted faith in technology and industrialization but also their attitudes toward other human beings and other living things. Economically secure yet dissatisfied with the status quo, these youthful rebels launched a massive push for social change that took a variety of forms, ranging from the civil rights movement, the anti-Vietnam War protests, and the large counterculture movement of the 1960s to the environmental movement that reached its peak in the 1970s.

The launching of *Sputnik* in 1957, the first manned space flight in 1961, and the first walk on the moon in 1969 provided the world with a whole new perspective on the place of the earth in the universe.

Commercial air travel had become commonplace in the 1950s, but by the early 1960s jets were replacing propeller planes on long-distance routes and changing relationships among people and businesses throughout the world. Computers were just starting to have an effect on the workplace, biotechnology was barely in the gestational stage, and the promises of science and technology had captured the public's imagination and loosened the government's purse strings.

Industrial and agricultural production were at an all-time high. Between 1950 and 1970, the increase in the production of goods and services was equal to the increase that had occurred between 1620, when the Pilgrims landed, and 1950.[1] The downside of this tremendous growth spurt was the enormous demand for energy and the subsequent increase in oil imports, the consumption of huge quantities of natural resources, and widespread industrial pollution. By the 1970s concern about the negative impacts of this growth had pushed environmental issues high on the national agenda.

## THE ENVIRONMENTAL MOVEMENT

Writers had been expounding on the problems of dwindling resources and pollution for decades, but they had left the general population unmoved and, in fact, uneducated. Then, in 1962, Rachel Carson, who had a substantial following from her works about the sea, produced *Silent Spring* [Document 99] and raised public consciousness about environmental degradation. The core message of Carson, Murray Bookchin [see Document 98], Paul Ehrlich [see Document 106], and other writers in the 1960s—that there is a connection between societal progress and environmental degradation, that human well-being is dependent on the well-being of the natural world, that the use of modern technology and chemicals may have serious negative as well as positive consequences, and that we as a nation could not continue to allow our natural resources to be destroyed or consumed at an ever increasing rate—was not new. For the past century, knowledgeable Americans had been discussing these issues. However, by stating their ideas in unambiguous and frequently alarmist terms, Carson and other environmentalists were able to arouse a very receptive public to grass-roots activism. As a legion of authors wrote elaborations on the silent spring theme, an environmental tsunami slowly built up into the modern environmental movement.

While the conservation movement at the turn of the century was initiated by a wealthy intelligentsia concerned about the future well-being of the nation, the new environmental movement was led and supported by ordinary people, together with segments of the scientific community, who recognized pollution as a source of immediate danger to their own health and well-being. In addition to being worried about the long-term effects of pesticides like DDT, people were also anxious about the imminent possibility of nuclear war. Despite the signing of the Nuclear Test Ban Treaty in 1963 [see Document 101] by signatory nations and other countries, ocean and underground nuclear testing persisted into the 1970s and beyond (the last U.S. nuclear test was conducted in 1992), and the United States and the Soviet Union continued to have huge stockpiles of nuclear weapons. Even in the late 1990s there has been some testing, carried out by nations that did not have nuclear capability in 1963.

Both the Union of Concerned Scientists (organized in 1969) and Greenpeace (founded in Canada in 1971, with an American branch organized the same year) were formed to confront the nuclear threat. As time passed, the two organizations broadened their focus from the advocacy of nuclear weapon destruction and the monitoring of nuclear testing to the control of radioactive and other toxic waste and to efforts to protect the quality of the environment.

Although the concept of environmentalism had entered the federal government's vocabulary toward the end of the Eisenhower administration [see Document 96], it was not until Richard Nixon's presidency that the federal government committed itself to environmental action. Nixon devoted a large part of his first State of the Union Address, in January 1970, to the need for an effective, and costly, environmental program [see Document 111A]. The impact of the environmental wave became evident in 1970, with Nixon's signing of the National Environmental Policy Act (NEPA) [Document 110] and the first Earth Day. Earth Day, April 22, 1970, marked a high point in popular concern about environmental issues and the beginning of a populist push for government regulation of activities and products that had a negative effect on the environment, and it served as a catalyst for a decade of environmental action on the local, national, and international levels. NEPA, probably the most far-reaching piece of federal environmental legislation ever passed, created the Environmental Protection Agency, which became the backbone of federal environmental action. During the next decade, strong environmental legislation was passed at every level of government. There were acts to ensure clean air [Document 112] and clean water [Document 115], to protect wetlands, scenic areas, wildlife habitats, forests, marine mammals, endangered species [Document 119], and human health, and to control the effects of agricultural chemicals and radioactive and other toxic wastes. In 1972 Oregon passed the first state bottle recycling law, marking the beginning of a nationwide effort to deal with the accumulating mounds of trash resulting from the mass distribution of packaged consumer goods and foods (including bottled soda, canned foods and drinks, and boxed cereals, whose availability had multiplied exponentially since the beginning of the century) and from the superabundance of printed matter (including newspapers, magazines, and tons of advertising matter).

Meanwhile, new environmental groups were forming and old conservation organizations, such as the Sierra Club, were refocusing and gaining new strength. Two of the most effective new organizations— the Natural Resources Defense Council and the Environmental Defense Fund [see Document 113]—took shape to support the causes of environmentalists in the courts. Both were outgrowths of local environmental groups with a much narrower mandate.

The post-Carson generation of writers sought to do more than just enact more effective environmental legislation. They wanted to change human attitudes as well as human behavior. In 1972 Justice William O. Douglas [see Document 117B] attempted to institutionalize Christopher Stone's advocacy of the rights of nature [see Document 116]. Stone's view of the relationship between humans and nature was a rethinking of a concept that had been stated over the centuries in various ways by such diverse writers as St. Francis, Thoreau, Muir, and Osborn [see Documents 44, 47, and 87], but in the Western world this had clearly been a minority viewpoint. However, in the 1970s, for the first time, large numbers of people began to question the idea that humans had a right or even an ability to dominate nature. The time had come to reexamine some of the country's core religious and philosophical values [see Document 105].

People also began to be concerned about the destabilization of ecosystems that had attained a balance over the course of thousands of years [see Documents 97 and 108]. Increasingly, they recognized that whenever someone or something upset this balance, there would be unforeseen and unpredictable consequences as the natural world reacted to the imbalance. Economic considerations alone were no longer sufficient justification for upsetting the balance [see Document 102].

The country's land development policies, to which there had been vocal opposition since the mid-nineteenth century [see Document 53], were increasingly attacked as a major cause of ecosystem destabilization. In the mid-1960s, as suburbs, shopping malls, and multilane highways spread across the country, there was mounting concern that too much open space was being paved over and that much of the construction was being done without any considerations for its impact on the environment. By 1965, California had passed a farmland conservation bill [see Document 103], and by 1970, the environment impact statement [see Document 109] had become part of the approval process for new land development projects.

## THE BEGINNING OF GLOBAL ENVIRONMENTAL CONCERN

In the 1960s and 1970s demographers and economists were again debating the old Malthusian arguments concerning the rate of population growth and the ability of the earth to feed increasing numbers of people [see Documents 106, 107, and 114], but the focus of these new discussions was on the global nature of the problem and on how world population growth could undermine the social, economic, and political stability of the United States as well as of other countries. Furthermore, it was also becoming obvious that people in Third World countries wanted to emulate the lifestyle depicted in American movies and television programs—a lifestyle based on high energy use and high material consumption.

The calling of the United Nations Conference on the Human Environment and the organization of the United Nations Environmental Programme (UNEP) in 1972 marked the beginning of a truly global approach to complex environmental issues [see Document 118].

## THE SEARCH FOR NEW SOLUTIONS

The mid-1970s were marked by inflation and unemployment (sparked in part by the Organization of Petroleum Exporting Countries' oil embargo in 1973). The war in Vietnam had come to an end and so had the era of protest. Although the oil crisis had brought home to millions of Americans the dangers of dependence on imported oil and had created interest in smaller cars and energy-saving technologies— President Jimmy Carter even had solar panels placed on the White House roof—the popular environmental focus was shifting from utopian solutions [see Document 120] to economically viable approaches that would enable Americans to maintain their way of life. While the general public was turning away from the idealists and from doomsayers such as Paul Ehrlich [see Document 106], some of the more radical environmental groups were borrowing the tactics of the protest movements of the 1960s and developing new tactics, ranging from civil disobedience to sabotage, to promote their political, ethical, and philosophical views and to effect changes in environmental policies [see Documents 121 and 122].

## DOCUMENT 96: The Surgeon General's Report on Environmental Health (1960)

By the end of the 1950s the effect on people's health and well-being of the tremendous post–World War II expansion of urbanization, industrialization, and population size was obvious. In its April 1959 report on appropriations for fiscal year 1960, the House of Representatives' Committee on Appropriations noted that "environmental factors affecting health have become increasingly significant" and recommended that "the Public Health Service make a thorough study of the environmental health problems and the most efficient organization of our facilities to meet these needs." The Surgeon General's report on "Environmental Health" was produced in response to that request. It underscored "the multiplying and far-reaching effects of new and complex problems in the field of environmental heath, and the immediate need for their identification and control." Consequently, in the 1961 budget of the Department of Health, Education, and Welfare, "environmental health activities" appeared as a separate line item for the first time.

It is not being overdramatic to suggest that threats from our environment, actual and potential, cannot only generate wholly undesirable effects on the health and well-being of isolated individuals, but under certain circumstances could affect large segments of our population and conceivably threaten the very existence of our Nation.

* * *

[O]ver the past several decades significant and growing nonbiological health hazards in the environment have arisen. Our increased use of materials and products is attended by increasing quantities of potentially toxic substances in our environment of limited physical dimensions. The uses of energy, accompanied by mechanization of processes and services, produce additional hazards of noise, other physical forces, and accidents. The diverse and growing beneficial uses of nuclear energy, potentially of enormous benefit, are producing a whole new spectrum of hazards to health for present and succeeding generations.

While the biological health hazard normally assaults the individual in discrete and separate instances, the chemical and physical hazards come more often in intermittent or continuous doses, which reach him in a variety of ways. It is the total and cumulative exposure of the individual to ionizing radiations which is now recognized to be important, no matter how the separate components reach him. Of no less concern is the total exposure to chemical toxicants, portions of which reach the individual separately through air, water, and food.

Social factors in the environment are contributing to the newer environmental health hazards. Increasing urbanization, the growth and coalescence of large metropolitan complexes, and new patterns of living compound the sources of environmental hazards and concentrate the people affected by them. Reactions between hazards and people are thus multiplied, and controls, to be effective, must be correspondingly more efficient.

The magnitude of our newer environmental health hazards is expanding at more than a linear rate. Acceleration arises from both population growth and technological advance. One of the most significant new elements in man's economic thinking is that the rate of economic growth can be accelerated by planned research and development. Accordingly, our expenditures for industrial research and development have been expanding impressively over the past decade. By comparison, expansion in research activities on the attendant health hazards has been substantially less.

Growing public concern about air and water pollution, food additives, and toxic residues and radiation exposure has, in some cases, led to pressures or demands for control actions for which rational scientific and technological procedures do not now exist.

To a large extent, the programs and activities for dealing with recognized environmental health problems have been developed and carried out each somewhat independent of the other, in response to a specific urgency.... Too often the important interrelated biological and social implications have not been given adequate attention.

... A unifying concept is needed if the approach to specific problems is to be most effective.

* * *

New chemicals, many of them with toxic properties or capabilities, are being produced and marketed, and put into use at a rapid rate. These include plastics, plasticizers, additives to fuels and foods, pesticides, detergents, abrasives.... It is estimated that 400–500 totally new chemicals are put into use each year. With many of these products, new waste byproducts are created which must be disposed of.

Although many commonly used chemicals are checked for toxicity, much is still unknown about their long-term potential hazards. The total dose of chemicals absorbed by man through numerous channels in minute and diluted amounts over his lifetime may be damaging his health.

\* \* \*

Water and air, food and milk, solid wastes, insects and rodents, and shelter are among the channels through which man's health may be endangered. Controls directed at these vehicles may eliminate or reduce not only known hazards, but unknown or little understood ones as well.

\* \* \*

Today, the modern supermarket and frozen food locker permit the use of a wide variety of foods, with resulting nutritional benefits. But modern methods of growing and processing foods introduce new hazards of pesticide spray residuals, preservatives and other food additives, and even contaminants related to packaging, which require attention for control.

*Source*: Surgeon General, "Environmental Health," Report to the House Committee on Appropriations (Washington, D.C.: Public Health Service, Department of Health, Education, and Welfare, January 1960), in *Departments of Labor and Health, Education, and Welfare Appropriations for 1961, Hearings before the Subcommittee on Appropriations*, House of Representatives, 86th Cong., 2nd sess. (Washington, D.C.: Government Printing Office, 1960), pp. 6, 7–8, 9–10.

## DOCUMENT 97: Lorus J. Milne and Margery Milne on the Balance of Nature (1960)

The late Lorus Milne, a professor of biology at the University of New Hampshire, and his wife Margery Milne, who also had obtained a Ph.D. in biology, were disturbed by large-scale efforts to eliminate pestiferous animals, insects, and plants. They recognized that even nuisance species prove useful and that their wholesale destruction would upset nature's precarious balance. The Milnes' review in the *New York Times* of Rachel Carson's *Silent Spring* [see Document 99] focused attention on another writer who was concerned about efforts to eliminate pests.

Often we forget the contrast between man's one-crop fields and nature's endless variety. One crop is the ultimate in imbalance, and must be defended constantly. We overlook, too, how many animals do both good and bad to man. The prairie dog's taste for foliage conflicts with ou[r] interests, but its liking for grasshoppers is a help. Nor is it easy to realize how small an area is home to many native animals. If man removes as weeds all except his crop plants, creatures with no special liking for the economic vegetation may eat it rather than travel to remote supplies of more palatable kinds. Other animals are driven off, upsetting still more the complex balance of undisturbed prairie.

Perhaps we tend to overlook our natural allies through unfamiliarity with modern situations where a true balance can be found. One of these was discovered recently, not on a prairie but in California's avocado orchards. So perfect is biological control in this cropland that each year less than one per cent of the total avocado acreage needs pest-control treatment. Yet insect enemies are present, unnoticed although ready to attack. An experiment tried near San Diego proved how quickly they can respond to opportunity.

For a period of eighty-four days, entomologists removed by hand every helpful parasitic and predatory animal they could find in one portion of a single tree. Within this trial period, caterpillars of one kind multiplied so rapidly that, to save the leaves, it was necessary to destroy the insects one by one. Two other avocado enemies throve until it seemed that similar action might have to be taken against them. But all of this change occurred only in the experimental portion of the tree. Other branches of the same avocado and other trees in the same grove presented no pest problem. Man's natural allies there continued their efficient control—at no cost to the growers.

On prairie ranch land, a balance can be reached if livestock are managed carefully. Prairie dogs cease to be a menace if a good cover of grass is present—neither too much nor too little. Too much grass is a sign of failure to utilize the land economically for cattle, although underuse of this kind can displace prairie dogs entirely. Too little grass leads to more prairie dogs, and often they are blamed unfairly for the barren ground produced by overgrazing by man's animals.

Fortunately, the ranchers are coming to understand the land they supervise. Some of them now recognize an obligation both to leave it in better condition than they found it, and to retain for the future the aesthetic values to be seen in native plants and animals. As this philosophy spreads, so too will a place for a few bison and prairie dogs. With the land once more in balance, man's own economy will also be in order.

*Source*: Lorus J. Milne and Margery Milne, *The Balance of Nature* (New York: Knopf, 1961), pp. 19–21.

## DOCUMENT 98: Murray Bookchin on the Synthetic Environment (1962)

Murray Bookchin is the author of numerous books on environmental and urban issues. Writing under the pen name Lewis Herber, he warned that the use of technology and technological innovations could have unanticipated effects and create new and unexpected environmental problems, often more serious than the problems the technology was intended to solve. Bookchin's book *Our Synthetic Environment* was published a few months earlier than Rachel Carson's *Silent Spring* and contained, in part, the same message, but it failed to capture as wide an audience.

The problems of our synthetic environment can be summed up by saying that nonhuman interests are superseding many of our responsibilities to human biological welfare. To a large extent, man is no longer working for himself. Many fields of knowledge and many practical endeavors that were once oriented toward the satisfaction of basic human wants have become ends in themselves, and to an ever-greater degree these new ends are conflicting with the requirements for human health. The needs of industrial plants are being placed before man's need for clean air; the disposal of industrial wastes has gained priority over the community's need for clean water. The most pernicious laws of the market place are given precedence over the most compelling laws of biology.

* * *

The problems created by our conflict with nature are dramatically exemplified by our chemical war against the insect world. During the past two decades, a large number of insecticides have been developed for general use on farms and in the home. The best-known and most widely used preparations are the chlorinated hydrocarbons, such as DDT, methoxychlor, dieldrin, and chlordane. The chlorinated hydrocarbons are sprayed over vast acres of forest land, range land, crop land, and even semi-urban land on which there are heavy infestations of insects. . . . Aside from the hazards that insecticides create for public health, many conservationists claim that extensive use of the new insecticides is impairing the ability of wildlife and beneficial insects to exercise control over pests. They point out that the insecticides are taking a heavy toll of life among fish, birds, small mammals, and useful insects. There is a great deal of evidence that the new chemicals are self-defeating. Not only have

they failed to eradicate most of the pests against which they are employed; in some cases, new pests and greater infestations have been created as a result of the damage inflicted on predators of species formerly
under control.

*Source*: Lewis Herber, *Our Synthetic Environment* (New York: Knopf, 1962), pp. 26,
53–54.

---

## DOCUMENT 99: Rachel Carson's *Silent Spring* (1962)

Rachel Carson, a marine biologist and author of three popular books
about the sea (including *The Sea Around Us*, which was on the best-
seller list for eighty-six weeks), was probably the one individual most
responsible for igniting public interest in pollution and other environmental issues. *Silent Spring*, which many consider the most important
book about the environment written in this century, caught the attention and the emotions of large numbers of people in the United States
and the rest of the world and alerted them to the dangers of man-made
poisons in the environment. The publication of the book marked the
beginning of popular concern about pollution and served as the starting
point for the environmental movement.

Carson became interested in the issue of pesticide spraying in the
mid-1940s. A neo-Malthusian, she worried that the need for increased
agricultural productivity—which would demand an increased use of
pesticides—would lead to a catastrophe. She was concerned that the
long-term use of synthetic pesticides could result in an accumulation
of toxins in part of the food chain and in the elimination of useful
insects. Carson never called for a complete ban on pesticides, but
rather appealed for their responsible use.

There was once a town in the heart of America where all life seemed
to live in harmony with its surroundings. The town lay in the midst of
a checkerboard of prosperous farms, with fields of grain and hillsides of
orchards where, in spring, white clouds of bloom drifted above the green
fields. In autumn, oak and maple and birch set up a blaze of color that
flamed and flickered across a backdrop of pines. Then foxes barked in
the hills and deer silently crossed the fields, half hidden in the mists of
the fall mornings.

Along the roads, laurel, viburnum and alder, great ferns and wild-
flowers delighted the traveler's eye through much of the year. Even in
winter the roadsides were places of beauty, where countless birds came
to feed on the berries and on the seed heads of the dried weeds rising

above the snow. The countryside was, in fact, famous for the abundance and variety of its bird life, and when the flood of migrants was pouring through in spring and fall people traveled from great distances to observe them. Others came to fish the streams, which flowed clear and cold out of the hills and contained shady pools where trout lay. So it had been from the days many years ago when the first settlers raised their houses, sank their wells, and built their barns.

Then a strange blight crept over the area and everything began to change. Some evil spell had settled on the community: mysterious maladies swept the flocks of chickens; the cattle and sheep sickened and died. Everywhere was a shadow of death. The farmers spoke of much illness among their families. In the town the doctors had become more and more puzzled by new kinds of sickness appearing among their patients. There had been several sudden and unexplained deaths, not only among adults but even among children, who would be stricken suddenly while at play and die within a few hours.

There was a strange stillness. The birds, for example—where had they gone? Many people spoke of them, puzzled and disturbed. The feeding stations in the backyards were deserted. The few birds seen anywhere were moribund; they trembled violently and could not fly. It was a spring without voices. On the mornings that had once throbbed with the dawn chorus of robins, catbirds, doves, jays, wrens, and scores of other bird voices there was now no sound; only silence lay over the fields and woods and marsh.

\* \* \*

As man proceeds toward his announced goal of the conquest of nature, he has written a depressing record of destruction, directed not only against the earth he inhabits but against the life that shares it with him. The history of the recent centuries has its black passages—the slaughter of the buffalo on the western plains, the massacre of the shorebirds by the market gunners, the near-extinction of the egrets for their plumage. Now, to these and others like them, we are adding a new chapter and a new kind of havoc—the direct killing of birds, mammals, fishes, and indeed practically every form of wildlife by chemical insecticides indiscriminately sprayed on the land.

\* \* \*

The question is whether any civilization can wage [such] relentless war on life without destroying itself, and without losing the right to be called civilized.

*Source*: Rachel Carson, *Silent Spring* (Boston: Houghton Mifflin, 1962), pp. 1–2, 85, 99.

---

## DOCUMENT 100: Stewart L. Udall on the Land Ethic (1963)

Stewart Udall came onto the national scene as a congressman from Arizona and later served as secretary of the interior in the Kennedy and Johnson administrations. In the mid-1970s, Udall worked valiantly to guide the Surface Mining Act (the strip-mining bill), which had twice been vetoed by President Gerald Ford, through the House of Representatives. The bill, finally passed in 1977, reclaimed several million acres of coal lands. In this selection he reminds readers of modern America's need for a land ethic like that proposed by Aldo Leopold in the 1940s [see Document 88].

Beyond all plans and programs, true conservation is ultimately something of the mind—an ideal of men who cherish their past and believe in their future. Our civilization will be measured by its fidelity to this ideal as surely as by its art and poetry and system of justice. In our perpetual search for abundance, beauty, and order we manifest both our love for the land and our sense of responsibility toward future generations.

Most Americans find it difficult to conceive a land ethic for tomorrow. The pastoral American of a century ago, whose conservation insights were undeveloped, has been succeeded by the asphalt American of the 1960's, who is shortsighted in other ways. Our sense of stewardship is uncertain partly because too many of us lack roots in the soil and the respect for resources that goes with such roots. Too many of us have mistaken material ease and comfort for the good life. Our growing dependence on machines has tended to mechanize our response to the world around us and has blunted our appreciation of the higher values.

\* \* \*

One of the paradoxes of American society is that while our economic standard of living has become the envy of the world, our environmental standard has steadily declined. We are better housed, better nourished, and better entertained, but we are not better prepared to inherit the earth or to carry on the pursuit of happiness.

A century ago we were a land-conscious, outdoor people: the American face was weather-beaten, our skills were muscular, and each family drew sustenance directly from the land. Now marvelous machines make our lives easier, but we are falling prey to the weaknesses of an indoor nation and the flabbiness of a sedentary society.

A land ethic for tomorrow should be as honest as Thoreau's *Walden*, and as comprehensive as the sensitive science of ecology. It should stress the oneness of our resources and the live-and-help-live logic of the great chain of life. If, in our haste to "progress," the economics of ecology are disregarded by citizens and policy makers alike, the result will be an ugly America. We cannot afford an America where expedience tramples upon esthetics and development decisions are made with an eye only on the present.

*Source*: Stewart Udall, *The Quiet Crisis* (New York: Holt, Rinehart and Winston, 1963), pp. vii–viii, 188–90.

---

## DOCUMENT 101: John F. Kennedy on the Nuclear Test Ban Treaty (1963)

During the early years of the cold war, the race between the United States and the Soviet Union to develop and build up their nuclear arsenals resulted in numerous nuclear weapons tests. From the start there was concern about the effects of radiation from nuclear testing, but it was not until 1959, when the relative nuclear position of the United States seemed sufficiently strong, that President Dwight Eisenhower proposed a treaty to limit nuclear testing. Four years of negotiations were required before President John F. Kennedy was able to transmit the treaty to the Senate. The Nuclear Test Ban Treaty of 1963, between the United States, the Soviet Union, and the United Kingdom, banned nuclear testing in the atmosphere, in outer space, and in water.

Concern about radioactive fallout from nuclear weapons testing was one of the driving forces behind the development of the environmental movement and provided the impetus for the formation of such organizations as the Union of Concerned Scientists and Greenpeace.

This treaty advances, though it does not assure, world peace; and it will inhibit, though it does not prohibit, the nuclear arms race.

- While it does not prohibit the United States and the Soviet Union from engaging in all nuclear tests, it will radically limit the testing in which both nations would otherwise engage.
- While it will not halt the production or reduce the existing stockpiles of nuclear weapons, it is a first step toward limiting the nuclear arms race.

- While it will not end the threat of nuclear war or outlaw the use of nuclear weapons, it can reduce world tensions, open a way to further agreements and thereby help to ease the threat of war.
- While it cannot wholly prevent the spread of nuclear arms to nations not now possessing them, it prohibits assistance to testing in these environments by others; it will be signed by many other potential testers; and it is thus an important opening wedge in our effort to "get the genie back in the bottle."

. . . The treaty will curb the pollution of our atmosphere. While it does not assure the world that it will be forever free from the fears and dangers of radioactive fallout from atmospheric tests, it will greatly reduce the numbers and dangers of such tests.

*Source:* "Special Message to the Senate on the Nuclear Test Ban Treaty. August 8, 1963," in *Public Papers of the Presidents of the United States: John F. Kennedy, 1963* (Washington, D.C.: Government Printing Office, 1964), pp. 622–23.

## DOCUMENT 102: *Scenic Hudson Preservation Conference v. Federal Power Commission* (1965)

In the early 1960s Consolidated Edison, the power company for New York City and Westchester County (just north of the city), proposed the creation of a huge reservoir a thousand feet above the Hudson River on Storm King Mountain. According to the proposal, river water would be pumped into the reservoir and stored there until more power was needed to deal with peak demands in the system. The water, as it was released back down into the Hudson, would turn a generating turbine.

After the Federal Power Commission (FPC) granted a license for this substantial project, the Scenic Hudson Preservation Conference (a group consisting of several conservation organizations and representatives of three Westchester towns, which had been formed in opposition to the project) asked the court to require the FPC to reconsider approval of the license application and examine alternatives to the project as well as other evidence that had been available before the project was approved but had been ignored.

The circuit court of appeal's decision broke new ground and is seen by many as the beginning of modern environmental law. The decision was significant for four reasons: it was the first case to elevate environmental factors to a position of equal consideration with economic factors; it was the first to require the government to consider alternatives in evaluating proposals for projects with negative environmental impact (a requirement established as law in the National Environmental

Policy Act of 1969 [Document 110]); it was among the first to require federal agencies to develop evidence relevant to the public interest, not simply weigh evidence presented by the applicant for license; and it was the first modern case to allow environmental groups with no economic interest in the issue to sue based on environmental issues.

The Court of Appeals, Hays, Circuit Judge, held, inter alia, that Federal Power Commission licensing order and subsequent related orders would be set aside for failure of commission to compile record sufficient to support its decision and because it ignored certain relevant factors and failed to make thorough study of possible alternatives, and matter would be remanded for its new proceedings which were required to include study of preservation of natural beauty and historic shrines and fisheries questions.

Order ["of the Federal Power Commission granting an intervener a license to construct a pumped storage hydroelectric project"] set aside and case remanded with directions.

\* \* \*

To be licensed by the [Federal Power] Commission a prospective project must meet the statutory test of being "best adapted to a comprehensive plan for developing a waterway." . . .

If the Commission is properly to discharge its duty in this regard, the record on which it bases its determination must be complete. The petitioners and the public at large have a right to demand this completeness. It is our view, and we find, that the Commission has failed to compile a record which is sufficient to support its decision. The Commission has ignored certain relevant factors and failed to make a thorough study of possible alternatives to the Storm King project. While the courts have no authority to concern themselves with the policies of the Commission, it is their duty to see to it that the Commission's decisions receive that careful consideration which the statute [Federal Power Act] contemplates.

\* \* \*

The Storm King project is to be located in an area of unique beauty and major historical significance. The highlands and gorge of the Hudson offer one of the finest pieces of river scenery in the world. The great German traveler Baedeker called it "finer than the Rhine." Petitioners' complaint that the Commission must take these factors into consideration in evaluating the Storm King project is justified by the history of the Federal Power Act.

* * *

Respondent argues that "petitioners do not have standing to obtain review" because they "make no claim of any personal economic injury resulting from the Commission's action."

* * *

In order to insure that the Federal Power Commission will adequately protect the public interest in the aesthetic, conservational, and recreational aspects of power development, those who by their activities and conduct have exhibited a special interest in such areas, must be held to be included in the class of "aggrieved" parties.

* * *

The "case" or "controversy" requirement of Article III section 2 of the Constitution does not require that an "aggrieved" or "adversely affected" party have a personal economic interest.

* * *

In this case, as in many others, the Commission has claimed to be the representative of the public interest. This role does not permit it to act as an umpire blandly calling balls and strikes for adversaries appearing before it; the right of the public must receive active and affirmative protection at the hands of the Commission.

*Source*: Scenic Hudson Preservation Conference, Town of Cortland, Town of Putnam Valley and Town of Yorktown, Petitioners v. Federal Power Commission, Respondent, and Consolidated Edison Company of New York, Inc., Intervener, Docket 29853, United States Court of Appeals, Second Circuit, 354 *Federal Reporter*, 2nd series, No. 106 (1965), pp. 608–9, 612, 613, 615, 620.

---

## DOCUMENT 103: California Land Conservation Act (1965) and Article XXVIII of the California Constitution (1966)

In the post-World War II years, farmland and open space began to disappear rapidly as a result of pressures from population growth, commercial expansion, and property tax increases. The California Land Conservation Act of 1965 (usually referred to as the Williamson Act) was one of the first legislative attempts to address the loss of open space to development. The implementation of the Williamson Act was made possible by the addition, in November 1966, of Article XXVIII to the California Constitution. Article XXVIII (now section 8 of Article XIII) provided tax incentives to discourage farmers and landowners from selling their land to developers because of financial pressures. As townships, counties, and states across the nation established their own land conservation programs, they followed California's lead in using financial incentives (which in time came to include such things as tax abatements and easements as well as the sale of development rights) as a major method of controlling suburban sprawl.

## A. The Act

The Legislature finds:

(a) That the preservation of a maximum amount of the limited supply of agricultural land is necessary to the conservation of the state's economic resources, and is necessary not only to the maintenance of the agricultural economy of the state, but also for the assurance of adequate, healthful and nutritious food for future residents of this state and nation.

(b) That the discouragement of premature and unnecessary conversion of prime agricultural land to urban uses is a matter of public interest and will be of benefit to urban dwellers themselves in that it will discourage discontiguous urban development patterns which unnecessarily increase the costs of community services to community residents.

(c) That in a rapidly urbanizing society agricultural lands have a definite public value as open space, and the preservation in agricultural production of such lands, the use of which may be limited under the provisions of this chapter, constitutes an important physical, social, esthetic and economic asset to existing or pending urban or metropolitan developments.

## B. California Constitution, Article XXVIII

SECTION 1. The people hereby declare that it is in the best interest of the state to maintain, preserve, conserve and otherwise continue in existence open space land for the production of food and fiber and to assure the use and enjoyment of natural resources and scenic beauty for the economic and social well-being of the state and its citizens. The people further declare that assessment practices must be so designed as to permit the continued availability of open space lands for these purposes, and it is the intent of this article to so provide.

SEC. 2. . . . [T]he Legislature may define open space lands and provide that when such lands are subject to enforceable restriction, as specified by the Legislature, to the use thereof solely for recreation, for the enjoyment of scenic beauty, for the use of natural resources, or for production of food or fiber, such lands shall be valued for assessment purposes on such basis as is consistent with such restriction and use.

*Source*: **A**. *Sessions Laws of American States and Territories: California 1960–1969* (Ann Arbor, MI: Xerox University Microfilms, n.d.), *1965 Regular Session*, p. 3378. **B**. California Constitution, in *Sessions Laws . . . : California 1960–1969*, fiche 273, *1967*, p. A111.

---

## DOCUMENT 104: Kenneth E. Boulding on the Spaceship Economy (1966)

Economist Kenneth Boulding believed that the United States needed to develop economic policies for the future based on minimizing both consumption and production, and that the object of its economic policies should be to maintain its stock of resources as much as possible. Boulding, however, was familiar with the concept of entropy (developed in 1824 by Nicholas Sadi Carnot, a French engineer and contemporary of Malthus), which posits that the amount of unavailable energy will always increase, and he understood that there will always be some loss of resources.

The closed earth of the future requires economic principles which are somewhat different from those of the open earth of the past. For the sake of picturesqueness, I am tempted to call the open economy the "cowboy economy," the cowboy being symbolic of the illimitable plains and also associated with reckless, exploitative, romantic, and violent behavior, which is characteristic of open societies. The closed economy of the future might similarly be called the "spaceman" economy, in which the earth has become a single spaceship, without unlimited reservoirs of anything, either for extraction or for pollution, and in which, therefore, man must find his place in a cyclical ecological system which is capable of continuous reproduction of material form even though it cannot escape having inputs of energy. The difference between the two types of economy becomes most apparent in the attitude towards consumption. In the cowboy economy, consumption is regarded as a good thing and production likewise; and the success of the economy is measured by the amount of the throughput from the "factors of production," a part of which, at any rate, is extracted from the reservoirs of raw materials and noneconomic objects, and another part of which is output into the reservoirs of pollution. If there are infinite reservoirs from which material can be obtained and into which effluvia can be deposited, then the throughput is at least a plausible measure of the success of the economy.

The gross national product is a rough measure of this total throughput. It should be possible, however, to distinguish that part of the GNP which is derived from exhaustible and that which is derived from reproducible resources, as well as that part of consumption which represents effluvia and that which represents input into the productive system again. . . .

By contrast, in the spaceman economy, throughput is by no means a desideratum, and is indeed to be regarded as something to be minimized rather than maximized. The essential measure of the success of the economy is not production and consumption at all, but the nature, extent, quality, and complexity of the total capital stock, including in this the state of the human bodies and minds included in the system. In the spaceman economy, what we are primarily concerned with is stock maintenance, and any technological change which results in the maintenance of a given total stock with a lessened throughput (that is, less production and consumption) is clearly a gain. This idea that both production and consumption are bad things rather than good things is very strange to economists, who have been obsessed with the income-flow concepts to the exclusion, almost, of capital-stock concepts.

*Source*: Kenneth E. Boulding, "The Economics of the Coming Spaceship Earth," in Henry Jarrett, ed., *Environmental Quality in a Growing Economy* (Baltimore: Johns Hopkins Press, 1966), pp. 9–10.

---

## DOCUMENT 105: Lynn White, Jr., on Western Religions and the Environmental Crisis (1966)

In a speech presented to the American Association for the Advancement of Science (AAAS) in 1966, Lynn White, Jr., a historian and expert on the development of technology, examined some of the human abuses of the environment and suggested that no solution to the ecological crisis could be achieved until people were willing to abandon their belief that nature exists only to serve humans. He directed Americans to reexamine their fundamental beliefs and the attitudes underlying their relationship to the environment, saying that Westerners should "find a new religion or abandon the old one."[2] More than anyone else, White has been responsible for the "greening" of Judeo-Christian thinking and for the development of ecotheology.

Since both *science* and *technology* are blessed words in our contemporary vocabulary, some may be happy at the notions first, that, viewed historically, modern science is an extrapolation of natural theology and, second, that modern technology is at least partly to be explained as an

Occidental, voluntarist realization of the Christian dogma of man's transcendence of, and rightful mastery over, nature. But, as we now recognize, somewhat over a century ago science and technology—hitherto quite separate activities—joined to give mankind powers which, to judge by many of the ecologic effects, are out of control. If so, Christianity bears a huge burden of guilt.

I personally doubt that disastrous ecologic backlash can be avoided simply by applying to our problems more science and more technology. Our science and technology have grown out of Christian attitudes toward man's relation to nature which are almost universally held not only by Christians and neo-Christians but also by those who fondly regard themselves as post-Christians. Despite Copernicus, all the cosmos rotates around our little globe. Despite Darwin, we are *not*, in our hearts, part of the natural process. We are superior to nature, contemptuous of it, willing to use it for our slightest whim.

* * *

The greatest spiritual revolutionary in Western history, Saint Francis, proposed what he thought was an alternative Christian view of nature and man's relation to it: he tried to substitute the idea of the equality of all creatures, including man, for the idea of man's limitless rule of creation. He failed. Both our present science and our present technology are so tinctured with orthodox Christian arrogance toward nature that no solution for our ecologic crisis can be expected from them alone. Since the roots of our trouble are so largely religious, the remedy must also be essentially religious, whether we call it that or not. We must rethink and refeel our nature and destiny. The profoundly religious, but heretical, sense of the primitive Franciscans for the spiritual autonomy of all parts of nature may point a direction. I propose Francis as a patron saint for ecologists.

*Source*: Excerpted with permission from Lynn White, Jr., "The Historical Roots of Our Environmental Crisis," *Science* 155, no. 3767 (March 10, 1967): 1206, 1207. Copyright 1967 American Association for the Advancement of Science.

---

## DOCUMENT 106: Paul Ehrlich's *The Population Bomb* (1968)

Paul Ehrlich, an instructor in entomology when he wrote *The Population Bomb*, is now a professor of population studies at Stanford University. Like Malthus in the nineteenth century, he uses the concept of population doubling time as a scare tactic to awaken people to the threat of an impeding population crisis. Although *The Population Bomb* proved effective in raising people's consciousness of the relationship between population growth, resource depletion, and environmental degradation, many of the predictions that Ehrlich made in the book turned out to be wrong. In 1977 Ehrlich and Julian Simon [see Document 126] made a bet about the future cost of acquiring various mineral resources, with Ehrlich betting the cost would go up and Simon betting it would go down as a result of improved technology. Twenty years later, when the wager was settled, Ehrlich was the loser.

Americans are beginning to realize that the underdeveloped countries of the world face an inevitable population-food crisis. Each year food production in these countries falls a bit further behind burgeoning population growth, and people go to bed a little bit hungrier. While there are temporary or local reversals of this trend, it now seems inevitable that it will continue to its logical conclusion: mass starvation. The rich may continue to get richer, but the more numerous poor are going to get poorer. Of these poor, a minimum of ten million people, most of them children, will starve to death during each year of the 1970s. But this is a mere handful compared to the numbers that will be starving before the end of the century. And it is now too late to take action to save many of those people.

However, most Americans are not aware that the U.S. and other developed countries also have a problem with overpopulation. Rather than suffering from food shortages, these countries show symptoms in the form of environmental deterioration and increased difficulty in obtaining resources to support their affluence.

... Perhaps the best way to impress you with numbers is to tell you about "doubling time"—the time necessary for the population to double in size.

It has been estimated that the human population of 8000 B.C. was about five million people, taking perhaps one million years to get there from two and a half million. The population did not reach 500 million until almost 10,000 years later—about 1650 A.D. This means it doubled roughly once every thousand years or so. It reached a billion people around 1850, doubling in some 200 years. It took only 80 years or so for the next doubling, as the population reached two billion around 1930. We have not completed the next doubling to four billion yet, but we now have well over three and a half billion people. The doubling time at present seems to be about 35 years. Quite a reduction in doubling times: 1,000,000 years, 1,000 years, 200 years, 80 years, 35 years. Perhaps the meaning of a doubling time of around 35 years is best brought home by a theoretical exercise. Let's examine what might happen on the absurd assumption that the population continued to double every 35 years into the indefinite future.

If growth continued at that rate for about 900 years, there would be some 60,000,000,000,000,000 people on the face of the earth. Sixty million billion people. This is about 100 persons for each square yard of the Earth's surface, land and sea.

*Source*: Paul R. Ehrlich, *The Population Bomb*, rev. ed. (Rivercity, MA: Rivercity Press, 1975; reprint of 1968 Ballantine Books ed.), pp. 3–4.

---

## DOCUMENT 107: Garrett Hardin on Controlling Access to the Commons (1968)

Every community has *commons*—resources such as parks, fresh air, waste disposal sites, water, and waterways to which all members of the community have equal access or right. As a community's population increases, the demand for use of these commons also increases. It is inevitable that if the population continues to increase, eventually the demand for the common resources will exceed the supply.

According to the human ecologist Garrett Hardin, the time has come to do away with unfettered human access to the commons in four areas: (1) food gathering, (2) waste disposal and pollution, (3) use of open space for pleasure and free expression (limits must be placed on the right to make noise, set up billboards, etc.), and (4) procreation at will. Limits on liberty, he believes, are better than total environmental ruin. Hardin has no faith in the ability of humans to prevent environmental degradation without the imposition of strict controls on human activity.

The tragedy of the commons develops in this way. Picture a pasture open to all. It is to be expected that each herdsman will try to keep as many cattle as possible on the commons. Such an arrangement may work reasonably satisfactorily for centuries because tribal wars, poaching, and disease keep the numbers of both man and beast well below the carrying capacity of the land. Finally, however, comes the day of reckoning, that is, the day when the long-desired goal of social stability becomes a reality. At this point, the inherent logic of the commons remorselessly generates tragedy.

As a rational being, each herdsman seeks to maximize his gain. Explicitly or implicitly, more or less consciously, he asks, "What is the utility *to me* of adding one more animal to my herd?" This utility has one negative and one positive component.

1) The positive component is a function of the increment of one animal. Since the herdsman receives all the proceeds from the sale of the additional animal, the positive utility is nearly +1.

2) The negative component is a function of the additional overgrazing created by one more animal. Since, however, the effects of overgrazing are shared by all the herdsmen, the negative utility for any particular decision-making herdsman is only a fraction of −1.

Adding together the component partial utilities, the rational herdsman concludes that the only sensible course for him to pursue is to add another animal to his herd. And another; and another. . . . But this is the conclusion reached by each and every rational herdsman sharing a commons. Therein is the tragedy. Each man is locked into a system that compels him to increase his herd without limit—in a world that is limited. Ruin is the destination toward which all men rush, each pursuing his own best interest in a society that believes in the freedom of the commons. Freedom in a commons brings ruin to all.

. . . [N]atural selection favors the forces of psychological denial. The individual benefits as an individual from his ability to deny the truth even though society as a whole, of which he is a part, suffers.

\* \* \*

The tragedy of the commons as a food basket is averted by private property, or something formally like it. But the air and waters surrounding us cannot readily be fenced, and so the tragedy of the commons as a cesspool must be prevented by different means, by coercive laws or taxing devices that make it cheaper for the polluter to treat his pollutants than to discharge them untreated. We have not progressed as far with the solution of this problem as we have with the first. Indeed, our particular concept of private property, which deters us from exhausting the positive resources of the earth, favors pollution. The owner of a factory on the bank of a stream—whose property extends to the middle of the stream—often has difficulty seeing why it is not his natural right to muddy the waters flowing past his door. . . .

The pollution problem is a consequence of population. It did not much matter how a lonely American frontiersman disposed of his waste. . . . But as population became denser, the natural chemical and biological recycling processes became overloaded, calling for a redefinition of property rights.

\* \* \*

The most important aspect of necessity that we must now recognize, is the necessity of abandoning the commons in breeding. No technical solution can rescue us from the misery of overpopulation. Freedom to breed will bring ruin to all. At the moment, to avoid hard decisions many of us are tempted to propagandize for conscience and responsible parenthood. The temptation must be resisted, because an appeal to independently acting consciences selects for the disappearance of all conscience in the long run, and an increase in anxiety in the short.

*Source*: Excerpted with permission from Garrett Hardin, "The Tragedy of the Commons," *Science* 162, no. 1243 (December 13, 1968): 88, 89, 92. Copyright 1968 American Association for the Advancement of Science.

---

## DOCUMENT 108: John Teal and Mildred Teal on the Productivity of the Salt Marsh (1969)

Between 1950 and 1980 the population of the coastal regions of the United States increased by more than 30 million people, fueled in part by the baby boom and by a tremendous growth in disposable income. In order to expand the land available for homes and agriculture, local wetlands were often filled in, with the first batch of landfill frequently consisting of garbage. As a result, not only were the barrier islands modified, but the salt marshes behind them—the wetlands—began to disappear. Estimates indicate that by the mid-1950s there was a loss of 25 percent of the coastal wetlands that had existed when Columbus reached America.

The call to protect these wetlands came in 1969, when John and Mildred Teal published *Life and Death of a Salt Marsh*. The Teals, who had spent years studying salt marshes, made the public aware that wetlands are not wastelands, as most Americans have long believed [see Document 19], but rather are some of the most productive areas in the world, and that much of the fish and other seafood we eat is dependent, directly or indirectly, on the existence of salt marshes.

Estuaries in general and salt marshes in particular are unusually productive places. None of the common agriculture, except possible rice and sugarcane production, comes close to producing as much potential animal food as do the salt marshes. The agricultural crops which approach this high figure are fertilized and cultivated at great expense. The marsh is fertilized and cultivated only by the tides.

Marshes are productive for several reasons, all of which are a result of the meeting of land and sea. The tides continually mix the waters and, by their rise and fall, water the plants. Harmful accumulations of waste products are diluted and removed. Nutrients are brought in continuous supply. The plants can put energy into growth, which in another environment they might have to use for collecting nutrients. The meeting of fresh and salt waters tends to trap nutrients in the regions of such meeting and this concentration of nutrients promotes plant growth. The materials coming down the rivers tend to focculate and settle out when they reach the increasingly salty estuarine waters. But the settled materials are not lost. *Spartina* roots use them. The tides continually stir the settled material and make the nutrients available for use by the phytoplankton.

Clams, oysters, and mussels help form the sediment and nutrient trap because of their method of feeding. They remove all of the particles from a large volume of water as it passes over their gills and deposit most of them in neat pseudofeces bundles.

\* \* \*

Nutrient concentrations in soils are thousands of times greater than those in waters. As a result, plants growing in soils are more productive than those floating in the water. A relatively small amount of soil can bring a large amount of plant nutrients into the marsh. It may even be deposited on just a few days of the year when the river carries an extra heavy burden of sediment resulting from an occasional flooding rain far from the marsh.

\* \* \*

Finally, the marshes are productive because there is almost no time during the year, even in the north, when there is not some plant growth taking place. In the south, it is warm enough for the *Spartina* to grow all year. At the latitudes of the mangroves, all plants grow all of the time if they are well watered. But in the north, where the land plants and *Spartina* cease activity during the winter, the algae in the marshes continue to grow throughout the year. The relatively constant growth of the mud algae in southern marshes and the midwinter blooming of the algae in the waters of the northern marshes are examples of year-round photosynthetic activity which helps move the marshes ahead of their neighboring terrestrial or marine areas in production.

*Source*: John and Mildred Teal, *Life and Death of a Salt Marsh* (New York: Ballantine Books, 1969), pp. 193–95.

## DOCUMENT 109: Ian McHarg on the Fitness of Ecosystems (1969)

Ian McHarg, the father of the environmental impact statement, which requires developers to justify the effects of their actions on the natural world, was ever conscious that natural landscapes had evolved over time because they were the most suitable for their ecosystems.

If evolutionary success is revealed by the existence of things and creatures, then their creative adaptations will be visible, not only in the organs and the organisms, but in ecosystems as well. If this is so, then the natural communities of plants and animals which the first colonists encountered in aboriginal America were the best expression of environmental adaptation, exploited by available organisms. Where these persist, this will hold true today. Thus not only is there an appropriate community of creatures for any environment, and successional stages towards this climax, but the community is, in fact, expressive of its appropriateness, its fitness.

This is a conclusion of enormous magnitude to those who are concerned with the land and its aspect: that there is a natural association which is most appropriate—indeed, in the absence of man, one which would be inevitable for every place upon the earth—and that that community of creatures is expressive of its fitness. This I would call the identity of the given form.

If this is so, then we can accept that within any generalized area there will be ideal examples, both of fitness and the expression of fitness. In these locations, presumably, there are some special successes that are visible and comprehensible. The ecosystem, the organisms and their organs are not only fit, but are most fitting. This is an important conception because it has a relevance to the man who wishes to design with nature. The man who seeks to create metaphysical symbols is really concerned with idealizing.

\* \* \*

[M]an is involved in exactly the same type of activity as the chambered nautilus, the bee and the coral, and subject to exactly the same tests of survival and evolution. Form is not the preoccupation of dilettantes but a central and indissoluble concern for all life.

Certainly we can dispose of the old canard, "form follows function." Form follows nothing—it is integral with all processes. Then form is indivisibly meaningful form, but it can reveal ill fit, misfit, unfit, fit and most fitting. There seems to be no good reason to change these criteria for human adaptations. Is the environment fit for man? Is the adaptation that is accomplished fit for the environment? Is the fit expressed in form?

*Source*: Ian McHarg, *Design with Nature* (Garden City, NY: Natural History Press/Doubleday, 1969), pp. 169–70, 173.

---

## DOCUMENT 110: National Environmental Policy Act (1969)

---

The National Environmental Policy Act (NEPA), signed into law by President Richard Nixon on January 1, 1970, set the stage for much of the environmental legislation passed in the 1970s. Its impact on economic and technological development in the United States has been broader than that of any other environmental legislation. Under this law, all federal agencies and departments are required to "monitor, evaluate, and control" their activities to protect and enhance the quality of the environment. The agencies must inform the public of their actions, review alternate courses of action, and allow the public to comment on those actions.

The law generated new power for the public and for organizations concerned with environmental issues. It also caused the creation of new divisions of law firms and corporations dealing in areas of the environment in which conflict might occur.

The environmental impact statement, the public document required by NEPA for any significant land development involving a federal agency, became the focal point for many environmental battles.

Sec. 101. (a) The Congress, recognizing the profound impact of man's activity on the interrelations of all components of the natural environment, particularly the profound influences of population growth, high-density urbanization, industrial expansion, resource exploitation, and new and expanding technological advances and recognizing further the critical importance of restoring and maintaining environmental quality to the overall welfare and development of man, declares that it is the continuing policy of the Federal Government, in cooperation with State and local governments, and other concerned public and private organizations, to use all practicable means and measures, including financial and technical assistance, in a manner calculated to foster and promote the general welfare, to create and maintain conditions under which man and nature can exist in productive harmony, and fulfill the social, economic, and other requirements of present and future generations of Americans.

(b) In order to carry out the policy set forth in this Act, it is the continuing responsibility of the Federal Government to use all practicable means, consistent with other essential considerations of national policy, to improve and coordinate Federal plans, functions, programs, and resources to the end that the Nation may—

(1) fulfill the responsibilities of each generation as trustee of the environment for succeeding generations;

(2) assure for all Americans safe, healthful, productive, and esthetically and culturally pleasing surroundings;

(3) attain the widest range of beneficial uses of the environment without degradation, risk to health or safety, or other undesirable and unintended consequences;

(4) preserve important historic, cultural, and natural aspects of our national heritage, and maintain, wherever possible, an environment which supports diversity and variety of individual choice;

(5) achieve a balance between population and resource use which will permit high standards of living and a wide sharing of life's amenities; and

(6) enhance the quality of renewable resources and approach the maximum attainable recycling of depletable resources.

(c) The Congress recognizes that each person should enjoy a healthful environment and that each person has a responsibility to contribute to the preservation and enhancement of the environment.

SEC. 102. The Congress authorizes and directs that, to the fullest extent possible: (1) the policies, regulations, and public laws of the United States shall be interpreted and administered in accordance with the policies set forth in this Act, and (2) all agencies of the Federal Government shall—

(A) utilize a systematic, interdisciplinary approach which will insure the integrated use of the natural and social sciences and the environmental design arts in planning and in decisionmaking which may have an impact on man's environment;

(B) identify and develop methods and procedures, in consultation with the Council on Environmental Quality [in the Executive Office of the President] established by subchapter II of this Act, which will insure that presently unquantified environmental amenities and values may be given appropriate consideration in decisionmaking along with economic and technical considerations;

(C) include in every recommendation or report on proposals for legislation and other major Federal actions significantly affecting the quality of the human environment, a detailed statement by the responsible official on—

(i)   the environmental impact of the proposed action,

(ii)  any adverse environmental effects which cannot be avoided should the proposal be implemented,

(iii) alternatives to the proposed action,

(iv)  the relationship between local short-term uses of man's environment and the maintenance and enhancement of long-term productivity, and

(v)   any irreversible and irretrievable commitments of resources which would be involved in the proposed action should it be implemented.
Prior to making any detailed statement, the responsible Federal official shall consult with and obtain the comments of any Federal agency which has jurisdiction by law or special expertise with respect to any environmental impact involved. Copies of such statement and the comments and views of the appropriate Federal, State, and local agencies, which are authorized to develop and enforce environmental standards, shall be made available to the President, the Council on Environmental Quality and to the public. . . .

(D) study, develop, and describe appropriate alternatives to recommended courses of action in any proposal which involves unresolved conflicts concerning alternative uses of available resources;

(E) recognize the worldwide and long-range character of environmental problems and, where consistent with the foreign policy of the United States, lend appropriate support to initiatives, resolutions, and programs designed to maximize international cooperation in anticipating and preventing a decline in the quality of mankind's world environment.

(F) make available to States, counties, municipalities, institutions, and individuals, advice and information useful in restoring, maintaining, and enhancing the quality of the environment;

(G) initiate and utilize ecological information in the planning and development of resource-oriented projects; and

(H) assist the Council on Environmental Quality.

*Source*: Public Law 91–190, *United States Statutes at Large*, Vol. 83 (Washington, D.C.: Government Printing Office, 1970), 91st Cong., 1st Sess., January 1, 1970, pp. 852–54.

## DOCUMENT 111: Richard Nixon on the Need for Environmental Regulation (1970)

Although few people would consider Richard Nixon to have been an environmentalist, he signed into law—albeit with reluctance on occasion—some of the most important environmental legislation of the twentieth century and supported the ratification of several major international environmental agreements. Much of his first State of the Union Address was devoted to a discussion of environmental issues.

Impetus for the amendments to strengthen the 1954 international Convention for the Prevention of the Pollution of the Sea by Oil, and for Nixon's and the general public's support of these amendments, came from a series of high-profile oil spill accidents, including the collision of the tanker *Torrey Canyon* in 1967 and the massive oil spill off the coast of Santa Barbara, California, in 1969.

### A. From the State of the Union Address, January 22, 1970

In the next 10 years we shall increase our wealth by 50 percent. The profound question is: Does this mean we will be 50 percent richer in a real sense, 50 percent better off, 50 percent happier?

Or does it mean that in the year 1980 the President standing in this place will look back on a decade in which 70 percent of our people lived in metropolitan areas choked by traffic, suffocated by smog, poisoned by water, deafened by noise, and terrorized by crime?

These are not the great questions that concern world leaders at summit conferences. But people do not live at the summit. They live in the foothills of everyday experience, and it is time for all of us to concern ourselves with the way real people live in real life.

The great question of the seventies is, shall we surrender to our surroundings, or shall we make our peace with nature and begin to make reparations for the damage we have done to our air, to our land, and to our water?

Restoring nature to its natural state is a cause beyond party and beyond factions. It has become a common cause of all the people of this country. It is a cause of particular concern to young Americans, because they more than we will reap the grim consequences of our failure to act on programs which are needed now if we are to prevent disaster later.

Clean air, clean water, open spaces—these should once again be the birthright of every American. If we act now, they can be.

We still think of air as free. But clean air is not free, and neither is clean water. The price tag on pollution control is high. Through our years of past carelessness we incurred a debt to nature, and now that debt is being called.

The program I shall propose to Congress will be the most comprehensive and costly program in this field in America's history.

. . .

I shall propose to this Congress a $10 billion nationwide clean waters program to put modern municipal waste treatment plants everywhere in America where they are needed to make our waters clean again. . . .

As our cities and suburbs relentlessly expand, those priceless open spaces needed for recreation areas accessible to their people are swallowed up—often forever. Unless we preserve these spaces while they are still available, we shall have none to preserve. . . .

The automobile is our worst polluter of the air. Adequate control requires further advances in engine design and fuel composition. We shall intensify our research, set increasingly strict standards, and strengthen enforcement standards—and we shall do it now.

We can no longer afford to consider air and water common property, free to be abused by anyone without regard to the consequences. Instead, we should begin now to treat them as scarce resources, which we are no more free to contaminate than we are free to throw garbage into our neighbor's yard.

This requires comprehensive new regulations. It also requires that, to the extent possible, the price of goods should be made to include the costs of producing and disposing of them without damage to the environment.

Now, I realize that the argument is often made that there is a fundamental contradiction between economic growth and the quality of life, so that to have one we must forsake the other.

The answer is not to abandon growth, but to redirect it. For example, we should turn toward ending congestion and eliminating smog the same reservoir of inventive genius that created them in the first place.

### B. From a Special Message to Congress on Marine Pollution from Oil Spills, May 20, 1970

The oil that fuels our industrial civilization can also foul our natural environment.

The threat of oil pollution from ships—both at sea and in our harbors—represents a growing danger to our marine environment. With the expansion of world trade over the past three decades, seaborne oil transport has multiplied tenfold and presently constitutes more than 60 percent of the world's ocean commerce.

This increase in shipping has increased the oil pollution hazard. Within the past ten years, there have been over 550 tanker collisions, four-fifths of which have involved ships entering or leaving ports. The routine discharge by tankers and other ships of oil and oily wastes as a part of their regular operation is also a major contributor to the oil pollution problem.

. . . The growing threat from oil spills can be contained—not by stopping industrial progress—but through a careful combination of international cooperation and national initiatives.

*Source: Public Papers of the Presidents of the United States: Richard Nixon, 1970* (Washington, D.C.: U.S. Government Printing Office, 1971), pp. 12–13, 443.

## DOCUMENT 112: Clean Air Act Amendments (1970)

The Clean Air Act of 1963, building on the Clean Air Act of 1955 [Document 93], authorized the Department of Health, Education, and Welfare to establish clean air criteria, but it left enforcement of these regulations to the states. A 1967 law required the states to develop implementation plans for the air quality goals they set.

In 1970 Congress took the next, very large step. The Clean Air Act Amendments of 1970 provided for the establishment of national air quality standards and required the states to develop implementation plans and set deadlines for meeting these standards. The law allowed the federal government to regulate, through the Environmental Protection Agency, emissions from automobiles and the composition of automobile fuels and additives. The Clean Air Act Amendments of 1970, together with the Clean Air Act Amendments passed in 1990, had enormous implications for industry, especially the automobile industry, and have had and will continue to have a huge effect on both the health and the economy of Americans.

### National Ambient Air Quality Standards

SEC. 109. (a) (1) The Administrator [the Environmental Protection Agency]—

(A) within 30 days after the enactment of the Clean Air Amendments of 1970, shall publish proposed regulations prescribing a national primary ambient air quality standard and a national secondary ambient air quality standard for each air pollutant for which air quality criteria have been issued prior to such date of enactment; and

(B) after a reasonable time for interested persons to submit written comments thereon (but no later than 90 days after the initial publication of such proposed standards) shall by regulation promulgate such proposed national primary and secondary ambient air quality standards with such modifications as he deems appropriate.

(2). With respect to any air pollutant for which air quality criteria are issued after the date of enactment of the Clean Air Amendments of 1970, the Administrator shall publish, simultaneously with the issuance of such criteria and information, proposed national primary and secondary ambient air quality standards for any such pollutant. . . .

(b) (1) National primary ambient air quality standards, prescribed under subsection (a), shall be ambient air quality standards the attainment and maintenance of which in the judgment of the Administrator, based on such criteria and allowing an adequate margin of safety, are requisite to protect the public health. . . .

(2) Any national secondary ambient air quality standard prescribed under subsection (a) shall specify a level of air quality the attainment and maintenance of which in the judgment of the Administrator, based on such criteria, is requisite to protect the public welfare from any known or anticipated adverse effects associated with the presence of such air pollutant in the ambient air. . . .

### Implementation Plans

Sec. 110. Each State shall, after reasonable notice and public hearings, adopt and submit to the Administrator, within nine months after the promulgation of a national primary ambient air quality standard (or any revision thereof) under section 109 for any air pollutant, a plan which provides for implementation, maintenance, and enforcement of such primary standard in each air quality control region (or portion thereof) within such State.

*Source*: Public Law 91–604, *U.S. Statutes at Large*, Vol. 84 (Washington, D.C.: Government Printing Office, 1971), 91st Cong., 2nd sess., Dec. 31, 1970, pp. 1679–80.

---

## DOCUMENT 113: Dennis Puleston on the Founding of the Environmental Defense Fund (1971)

---

Dennis Puleston, the chairman of the board of trustees of the Environmental Defense Fund, became an active environmentalist when he realized that the osprey were disappearing from Long Island, New York, where he lives. As often happens, concern about a local environmental problem led to involvement with much broader environmental issues.

Coal miners working deep in the earth learned long ago that one of their most treacherous enemies was the colorless, odorless gas methane. They also found that certain living creatures were more sensitive to this gas than they were, and thus could serve as an early warning system. So they took a canary down with them, and as long as the bird sang cheerfully they knew all was well. But if it became silent they watched it carefully, and if it rolled over dead the alarm rang out and they made a dash for the safety of the pit head.

In recent years many of us who are concerned with the condition of our environment have developed our own canaries. Mine happens to be a large, regal looking brown and white bird with strong, hooked talons and raptorial beak. . . . I can well recall my first visit to Gardiner's Island in 1948, when there were more than 300 active [osprey] nests, producing an average of more than two fledglings per nest. . . . [S]ubsequent frequent visits there convinced me that numbers were declining. By the mid-fifties the number of active nests and also the number of young per active nest were dropping steadily, and by the early sixties the drop had reached collapse proportion. On the mainland [of Long Island] the situation was even more serious; the osprey became a rare sight in places where formerly they had been abundant, and few successful nestings could be recorded. . . . [O]ne fact is sure; the osprey is an endangered species in the Northeast. . . . And whether we care for the osprey or not, surely its tragic situation is a warning of something that is of concern to our own health and welfare. Perhaps the canary is not yet dead, but it has definitely stopped singing.

Naturally, I wished to determine the reason for this decline. Factors such as human and animal disturbances and shortage of food did not apply here; something else was causing the lack of reproductive success. Then, in the early sixties, Rachel Carson's powerful and eloquent indictment of the chlorinated hydrocarbon pesticides burst into print [see Document 99], a book that some critics claim will be recorded as having a greater dramatic effect than almost any other literary work of the century. Several unhatched eggs that I brought back from osprey nests on Gardiner's Island confirmed her findings. Gas chromatography revealed the presence of high concentrations of DDT and its metabolites; certainly >13 parts per million (ppm) was sufficient to kill an embryo.

* * *

Although some of the environmental problems resulting from the widespread use of DDT were not yet known to us by the mid-sixties, we had sufficient evidence to convince us that we must act to bring about a curtailment in its use. Not only the reproductive failure of ospreys, but the almost total disappearance of the blueclaw crab in our bays, the diminished populations of reptiles, amphibians, many species of birds, butterflies, and honeybees—all were sounding the alarm. The canary was far from well.

It was not difficult to point a finger at the chief culprit. For the past 20 years, the Suffolk County Mosquito Control Commission had been aerially spraying DDT in its largely unsuccessful efforts to reduce mosquito populations. From a salt marsh at the mouth of the Carmans River in Brookhaven, mud samples were found with concentrations as high as 32 lbs/acre—this in an area far from any agricultural spraying. Appeals to the Commission had no effect; their job, they claimed, was to kill mosquitoes, and DDT was the best way to do it. Effects on other wildlife forms were not their concern. Besides, they insisted, wasn't DDT "harmless to animals"?

Thus the idea of court action was born. A local lawyer was anxious to try his mettle, and there was plenty of scientific testimony to support his case. This was provided by a small group of life scientists and conservationists, members of the local Brookhaven Town Natural Resources Committee. Affidavits, charts, photographs, and a substantial technical appendix were prepared and in early 1966 an Order to Show Cause, together with a plaintiff's affidavit, summons, and verified complaint were filed with the Supreme Court of the State of New York in Riverhead, Suffolk County. . . . In August the Court determined that ". . . upon all the facts before the Court . . . sufficient grounds exist for the discretion vested in the Court to stay a practice injurious to the County and its residents."

In November 1966, the action was heard. This was a "class action," whereby the plaintiff, instead of seeking personal damages, seeks to persuade the Court that, as a matter of equity and constitutional privilege, the citizenry has a right to the cleanest possible environment consistent with the general welfare. While such a right is not specifically mentioned in the Constitution, it can readily be inferred. This doctrine holds that those who pollute and disturb the environment adversely in the name of progress and economic necessity should be required to show that their actions are in the public interest. Thus, the suit heard in the Riverhead courtroom was a pioneering effort to establish vital precedents in future conservation law.

The six-day trial was taken up largely with the testimony of scientists and other expert witnesses presented by the plaintiff's legal counsel. The counsel for the defense was hard put to summon any valid scientific support for the necessity of using DDT for mosquito control or to show that it was harmless to nontarget organisms. In fact, any economic arguments in favor of DDT were nonexistent.

The court-imposed temporary ban on DDT use by the defendant held for about a year, at which time the judge ruled that, although it had been proven that DDT was contaminating the environment, action for a permanent ban must come from the state legislature. Victory, however, was snatched from an ultimate defeat, for orders from the County administration came in time to make the ban permanent.

* * *

Emboldened by the small but significant victory in Suffolk, the original group decided to follow their courtroom tactics on a broader scale. . . .

Accordingly, in the fall of 1967, the Environmental Defense Fund (EDF) was formed.

*Source*: Dennis Puleston, "Defending the Environment: A Case History," *Brookhaven Lecture Series*, No. 104, September 15, 1971 (Springfield, VA: National Technical Information Service, U.S. Department of Commerce, December 1971), pp. 1, 3, 4, 5.

## DOCUMENT 114: Barry Commoner on Nature, Man, and Technology (1971)

Barry Commoner's concern with the environment began with his alarm about nuclear proliferation but broadened into an understanding of the need for a new view of the human role in the environment. One of the first scientists to comprehend the relationship between political activism and social change, Commoner played a role in the formation of the Union of Concerned Scientists. As a public advocate and articulate spokesperson for a wide-ranging environmental, social, and political agenda, Commoner moved from an early career as a lecturer in biology at Washington University to presidential candidate in 1980.

Earth Week and the accompanying outburst of publicity, preaching, and prognostication surprised most people, including those of us who had worked for years to generate public recognition of the environmental crisis. What surprised me most were the numerous, confident explanations of the cause and cure of the crisis. For having spent some years in the effort simply to detect and describe the growing list of environmental problems—radioactive fallout, air and water pollution, the deterioration of the soil—and in tracing some of their links to social and political processes, the identification of a single cause and cure seemed a rather bold step. During Earth Week, I discovered that such reticence was far behind the times.

After the excitement of Earth Week, I tried to find some meaning in the welter of contradictory advice that it produced. It seemed to me that the confusion of Earth Week was a sign that the situation was so complex and ambiguous that people could read into it what ever conclusion their own beliefs—about human nature, economics, and politics—suggested. . . .

Earth Week convinced me of the urgency of a deeper public understanding of the origins of the environmental crisis and its possible
cures. . . .

Such an understanding must begin at the source of life itself: the
earth's thin skin of air, water, and soil, and the radiant solar fire that
bathes it. Here, several billion years ago, life appeared and was nourished by the earth's substance. As it grew, life evolved, its old forms
transforming the earth's skin and new ones adapting to these changes.
Living things multiplied in number, variety, and habitat until they
formed a global network, becoming deftly enmeshed in the surroundings
they had themselves created. This is the *ecosphere*, the home that life has
built for itself on the planet's outer surface.

Any living thing that hopes to live on the earth must fit into the ecosphere or perish. The environmental crisis is a sign that the finely
sculpted fit between life and its surroundings has begun to corrode. As
the links between one living thing and another, and between all of them
and their surroundings, begin to break down, the dynamic interactions
that sustain the whole have begun to falter and, in some places, stop.

* * *

Understanding the ecosphere comes hard because, to the modern
mind, it is a curiously foreign place. We have become accustomed to
think of separate, singular events, each dependent upon a unique, singular cause. But in the ecosphere every effect is also a cause: an animal's
waste becomes food for soil bacteria; what bacteria excrete nourishes
plants; animals eat the plants. Such ecological cycles are hard to fit into
human experience in the age of technology, where machine A always
yields product B, and product B, once used, is cast away, having no
further meaning for the machine, the product, or the user.

Here is the first great fault in the life of man in the ecosphere. We
have broken out of the circle of life, converting its endless cycles into
man-made, linear events: oil is taken from the ground, distilled into fuel,
burned in an engine, converted thereby into noxious fumes, which are
emitted into the air. At the end of the line is smog. Other man-made
breaks in the ecosphere's cycles spew out toxic chemicals, sewage, heaps
of rubbish—testimony to our power to tear the ecological fabric that has,
for millions of years, sustained the planet's life.

Suddenly we have discovered what we should have known long before: that the ecosphere sustains people and everything that they do; that
anything that fails to fit into the ecosphere is a threat to its finely balanced cycles; that wastes are not only unpleasant, not only toxic, but,
more meaningfully, evidence that the ecosphere is being driven towards
collapse.

*Source*: Barry Commoner, *The Closing Circle: Nature, Man, and Technology* (New York: Knopf, 1971), pp. 10–12.

## DOCUMENT 115: Clean Water Act (1972)

The Clean Water Act of 1972, amending the Federal Water Pollution Control Act of 1948, gave teeth to earlier legislation aimed at preventing or reducing water pollution and laid the groundwork for future efforts to preserve the nation's wetlands.

SEC. 101. The objective of this Act is to restore and maintain the chemical, physical, and biological integrity of the Nation's waters. In order to achieve this objective it is hereby declared that, consistent with the provisions of this Act—

(1) it is the national goal that the discharge of pollutants into navigable waters be eliminated by 1985;

(2) it is the national goal that wherever attainable, an interim goal of water quality which provides for the protection and propagation of fish, shellfish, and wildlife and provides for recreation in and on the water be achieved by July 1, 1983;

(3) it is the national policy that the discharge of toxic pollutants in toxic amounts be prohibited;

(4) it is the national policy that Federal financial assistance be provided to construct publicly owned waste treatment works;

(5) it is the national policy that areawide waste treatment management planning processes be developed and implemented to assure adequate control of sources of pollution in each State; and

(6) it is the national policy that a major research and demonstration effort be made to develop technology necessary to eliminate the discharge of pollutants into the navigable waters, waters of the contiguous zone, and the oceans.

SEC. 404. (a) The Secretary of the Amy, acting through the Chief of Engineers, may issue permits, after notice and opportunity for public hearings for the discharge of dredged or fill material into the navigable waters at specified disposal sites.

(b) Subject to subsection (c) of this section, each such disposal site shall be specified for each such permit by the Secretary of the Army (1) through the application of guidelines developed by the Administrator, in conjunction with the Secretary of the Army . . . , and (2) in any case where such guidelines under clause (1) alone would prohibit the specification of a site, through the application additionally of the economic impact of the site on navigation and anchorage.

(c) The Administrator is authorized to prohibit the specification (including the withdrawal of specification) of any defined area as a disposal site, and he is authorized to deny or restrict the use of any defined area for specification (including the withdrawal of specification) as a disposal site, whenever he determines, after notice and opportunity for public hearings, that the discharge of such materials into such area will have an unacceptable adverse effect on municipal water supplies, shellfish beds, and fishery areas (including spawning and breeding areas), wildlife, or recreational areas.

*Source*: Public Law 92–500, *United States Statutes at Large*, Vol. 86 (Washington, D.C.: Government Printing Office, 1973), 92nd Cong., 2nd sess., Oct. 18, 1972, pp. 816, 884.

---

## DOCUMENT 116: Christopher Stone Proposes Legal Rights for Natural Objects (1972)

Henry David Thoreau suggested that all living things have inherent rights that humans ought to recognize [see Document 44]. John Muir, in his early musings, revealed an inclination to the same line of thought [see Document 47] but later, recognizing political realities, turned to the usefulness of the wilderness to man as the basis for arguing for its preservation. In his essay "Should Trees Have Standing?" Christopher Stone, a lawyer, explored a similar but, in a way, more practical approach to acknowledging the rights of natural objects, and suggested that trees or forests or rivers or ecosystems that are about to be or have been damaged should be represented—as children, incompetent adults, corporations, and nations are—by concerned and appropriate guardians.

It is not inevitable, nor is it wise, that natural objects should have no rights to seek redress in their own behalf. It is no answer to say that streams and forests cannot have standing because streams and forests cannot speak. Corporations cannot speak either; nor can states, estates, infants, incompetents, municipalities or universities. Lawyers speak for them. . . . One ought, I think, to handle the legal problems of natural objects as one does the problems of legal incompetents—human beings who have become vegetables. . . .

On a parity of reasoning, we should have a system in which, when a friend of a natural object perceives it to be endangered, he can apply to a court for the creation of a guardianship. . . .

. . . If, for example, the Environmental Defense Fund should have reason to believe that some company's strip mining operations might be irreparably destroying the ecological balance of large tracts of land, it could, under this procedure, apply to the court in which the lands were situated to be appointed guardian. As guardian, it might be given rights of inspection (or visitation) to determine and bring to the court's attention a fuller finding on the land's condition. If there were indications that under the substantive law some redress might be available on the land's behalf, then the guardian would be entitled to raise the land's rights in the land's name, *i.e.*, without having to make the roundabout and often unavailing demonstration . . . that the "rights" of the club's members were being invaded. . . .

As far as adjudicating the merits of a controversy is concerned, there is also a good case to be made for taking into account harm to the environment—in its own right. . . . [T]he traditional way of deciding whether to issue injunctions in law suits affecting the environment, at least where communal property is involved, has been to strike some sort of balance regarding the economic hardships on *human beings*. . . .

The argument for "personifying" the environment, from the point of damage calculations, can best be demonstrated from the welfare economics position. Every well-working legal-economic system should be so structured as to confront each of us with the full costs that our activities are imposing on society. Ideally, a paper-mill, in deciding what to produce—and where, and by what methods—ought to be forced to take into account not only the lumber, acid and labor that its production "takes" from other uses in the society, but also what costs alternative production plans will impose on society through pollution.

*Source*: Christopher Stone, "Should Trees Have Standing? Toward Legal Rights for Natural Objects," *Southern California Law Review* 45 (1972): 450–501.

---

## DOCUMENT 117: *Sierra Club v. Morton* (1972)

Not long after Christopher Stone published his essay on the legal rights of natural objects [Document 116], William O. Douglas reinforced the idea in a dissenting opinion in the case *Sierra Club v. [Interior Secretary Rogers] Morton*. Douglas, who served as an associate justice on the highest court for more than thirty-five years, was an avid naturalist and aggressive advocate for wilderness preservation.

## A. The Supreme Court Decision

Action by membership corporation [Sierra Club] for declaratory judgment that construction of proposed ski resort and recreation area in national game refuge and forest would contravene federal laws and for preliminary and permanent injunctions restraining federal officials from approving or issuing permits for the project. . . . The Supreme Court, Mr. Justice Stewart [author of the affirming opinion], held that, in absence of allegation that corporation or its members would be affected in any of their activities or pastimes by the proposed project, the corporation, which claimed special interest in conservation of natural game refuges and forests, lacked standing under Administrative Procedure Act to maintain the action.

## B. William O. Douglas's Minority Opinion

The critical question of "standing" would be simplified and also put neatly in focus if we fashioned a federal rule that allowed environmental issues to be litigated before federal agencies or federal courts in the name of the inanimate object about to be despoiled, defaced, or invaded by roads and bulldozers and where injury is the subject of public outrage. Contemporary public concern for protecting nature's ecological equilibrium should lead to the conferral of standing upon environmental objects to sue for their own preservation. See Stone, Should Trees Have Standing?—Toward Legal Rights for Natural Objects [Document 116]. This suit would therefore be more properly labeled as Mineral King v. Morton.

Inanimate objects are sometimes parties in litigation. . . . The ordinary corporation is a "person" for purposes of the adjudicatory processes, whether it represents proprietary, spiritual, aesthetic, or charitable causes.

. . .

Mineral King is doubtless like other wonders of the Sierra Nevada such as Tuolumne Meadows and the John Muir Trail. Those who hike it, fish it, hunt it, camp in it, frequent it, or visit it merely to sit in solitude and wonderment are legitimate spokesmen for it, whether they may be few or many. Those who have that intimate relation with the inanimate object about to be injured, polluted, or otherwise despoiled are its legitimate spokesmen.

. . . [T]he problem is to make certain that the inanimate objects, which are the very core of America's beauty, have spokesmen before they are destroyed. It is, of course, true that most of them are under the control of a federal or state agency. The standards given those agencies are usually expressed in terms of the "public interest." Yet "public interest" has so many differing shades of meaning as to be quite meaningless on the environmental front. Congress accordingly has adopted ecological standards in the National Environmental Policy Act of 1969 . . . and guidelines for agency action have been provided by the Council on Environmental Quality. . . .

Yet the pressures on agencies for favorable action one way or the other are enormous. The suggestion that Congress can stop action which is undesirable is true in theory; yet even Congress is too remote to give meaningful direction and its machinery is too ponderous to use very often. The federal agencies of which I speak are not venal or corrupt. But they are notoriously under the control of powerful interests who manipulate them through advisory committees, or friendly working relations, or who have that natural affinity with the agency which in time develops between the regulator and the regulated. As early as 1894, Attorney General [Richard] Olney predicted that regulatory agencies might become "industry-minded." . . .

Years later a court of appeals observed "the recurring question which has plagued public regulation of industry [is] whether the regulatory agency is unduly oriented toward the interests of the industry it is designed to regulate, rather than the public interest it is designed to protect." . . .

The Forest Service—one of the federal agencies behind the scheme to despoil Mineral King—has been notorious for its alignment with lumber companies, although its mandate from Congress directs it to consider the various aspects of multiple use in its supervision of the national forests.

The voice of the inanimate object, therefore, should not be stilled. That does not mean that the judiciary takes over the managerial functions from the federal agency. It merely means that before these priceless bits of Americana (such as a valley, an alpine meadow, a river, or a lake) are forever lost or are so transformed as to be reduced to the eventual rubble of our urban environment, the voice of the existing beneficiaries of these environmental wonders should be heard.

Perhaps they will not win. Perhaps the bulldozers of "progress" will plow under all the aesthetic wonders of this beautiful land. That is not the present question. The sole question is who has standing to be heard?

Those who hike the Appalachian Trail into Sunfish Pond, New Jersey, and camp or sleep there, or run the Allagash in Maine, or climb the Guadalupes in West Texas, or who canoe and portage the Quetico Superior in Minnesota, certainly should have standing to defend those natural wonders before courts or agencies, though they live 3,000 miles away. Those who merely are caught up in environmental news or propaganda and flock to defend these waters or areas may be treated differently. That is why these environmental issues should be tendered by the inanimate object itself. Then there will be assurances that all of the forms of life which it represents will stand before the court—the pileated woodpecker as well as the coyote and bear, the lemmings as well as the trout in the streams. Those inarticulate members of the ecological group cannot speak. But those people who have so frequented the place as to know its values and wonders will be able to speak for the entire ecological community.

*Source: Supreme Court Reporter*, 768; 92 Supreme Court 1361 (1972), pp. 1361, 1369–71, 1374–75.

---

## DOCUMENT 118: Stockholm Declaration on the Human Environment (1972)

---

The United States was not alone in its awakening to environmental threats. In June 1972, in response to the expanding international environmental consciousness, the United Nations sponsored the Conference on the Human Environment in Stockholm. The declaration produced by the conference took note of the "dangerous levels of pollution in water, air, earth and living beings; major and undesirable disturbances to the ecological balance of the biosphere; destruction and depletion of irreplaceable resources; and gross deficiencies harmful to the physical, mental and social health of man, in the man-made environment."

The conference, which marked the beginning of the United Nations Environmental Programme (UNEP), recognized both the increasing demand by Third World countries for a Western-style standard of living and the desire of the industrialized nations for continued economic growth. The declaration that it produced underscored the fact that international environmental policy is clearly subordinate to the economic interests of individual states, as noted in principles 21 and 24. Nevertheless, during the more than a quarter century since its founding, UNEP has succeeded in encouraging the nations of the world to take major steps to decrease sea and air pollution. Indeed, many of the UNEP-sponsored protocols and conventions relating to the environment have been incorporated into U.S. law, although the U.S. Senate has balked at ratifying international environmental agreements that it views as detrimental to U.S. strategic or economic interests [see, for example, Documents 129 and 142B].

The Stockholm Declaration also committed the U.N. to the concept of *sustainable development*—the idea that, with suitable resource management, the resource base of nations, including soil, fisheries, water supplies, and forests, could be maintained, and that at the same time both underdeveloped and developed countries could continue to grow economically.

2. The natural resources of the earth including the air, water, land, flora and fauna and especially representative samples of natural ecosystems must be safeguarded for the benefit of present and future generations through careful planning or management, as appropriate.

3. The capacity of the earth to produce vital resources must be maintained and, wherever practicable, restored or improved.

4. Man has a special responsibility to safeguard and wisely manage the heritage of wildlife and its habitat which are now greatly imperiled by a combination of adverse factors. Nature conservation including wildlife must therefore receive importance for planning in economic development.

5. The non-renewable resources of the earth must be employed in such a way as to guard against the danger of their future exhaustion and to ensure that benefits from such employment are shared by all mankind.

6. The discharge of toxic substances or of other substances and the release of heat, in such quantities or concentrations as to exceed the capacity of the environment to render them harmless, must be halted in order to ensure that serious or irreversible damage is not inflicted upon ecosystems. . . .

7. States shall take all possible steps to prevent pollution of the seas by substances that are liable to create hazards to human health, to harm living resources and marine life, to damage amenities or to interfere with other legitimate uses of the sea.

21. States have, in accordance with the Charter of the United Nations and the principles of international law, the sovereign right to exploit their own resources pursuant to their own environmental policies, and the responsibility to ensure that activities within their jurisdiction or control do not cause damage to the environment of other States or of areas beyond the limits of national jurisdiction.

24. International matters concerning the protection and improvement of the environment should be handled in a cooperative spirit by all countries, big or small, on an equal footing. Cooperation through multilateral or bilateral arrangements or other appropriate means is essential to effectively control, prevent, reduce and eliminated adverse environmental effects resulting from activities conducted in all spheres, in such a way that due account is taken of sovereignty and interests of all States.

*Source: Stockholm Declaration on the Human Environment* (New York: United Nations Environmental Programme, 1972), pp. 1–4.

## DOCUMENT 119: Endangered Species Act (1973)

The Endangered Species Act, in its original form, came close to ac-knowledging the rights of all living things to exist. In the framing of the law, no references were made to the utility of the species; the task was only to identify and list them and to find a way to develop a plan to encourage their recovery.

The history of the act and its enforcement is filled with conflicts in which the rights of property owners and other financially interested parties are pitted against spokespeople for the preservation of endan-gered species. The conflict over the Tennessee Valley Authority's Tel-lico Dam, situated on the Little Tennessee River, and the endangered snail darter (a rare species of perch) has been cited for many years as the exemplification of the absurdity of some of these conflicts.[3] More recently, the controversy surrounding the preservation of the spotted owl in the ancient forests of the Northwest has produced a similar exchange of invectives. Over the years, compromises had to be made to enable the country to live with the act, and consequently the law has been substantially altered from the original version presented here.

The act has had its share of both successes and failures. Ospreys, brown pelicans, bald eagles, and peregrine falcons have moved from the brink of extinction to a certainty of survival if current environmental conditions continue.

(a) FINDINGS.—The Congress finds and declares that—

(1) various species of fish, wildlife, and plants in the United States have been rendered extinct as a consequence of economic growth and development untempered by adequate concern and conservation;

(2) other species of fish, wildlife, and plants have been so depleted in numbers that they are in danger of or threatened with extinction;

(3) these species of fish, wildlife, and plants are of esthetic, ecological, educational, historical, recreational, and scientific value to the Nation and its people;

(4) the United States has pledged itself as a sovereign state in the international community to conserve to the extent practicable the various species of fish or wildlife and plants facing extinction, pursuant to—

(A) migratory bird treaties with Canada and Mexico;

(B) the Migratory and Endangered Bird Treaty with Japan;

(C) the Convention on Nature Protection and Wildlife Preservation for the Western Hemisphere;

(D) the International Convention for the Northwest Atlantic Fisheries;

(E) the International Convention for the High Seas Fisheries of the North Pacific Ocean;

(F) the Convention on International Trade in Endangered Species of Wild Fauna and Flora; and

(G) other international agreements.

(5) encouraging the States and other interested parties, through Federal financial assistance and a system of incentives, to develop and maintain conservation programs which meet national and international standards is a key to meeting the Nation's international commitments and to better safeguarding, for the benefit of all citizens, the Nation's heritage in fish, wildlife, and plants.

(b) PURPOSES.—The purposes of this Act are to provide a means whereby the ecosystems upon which endangered species and threatened species depend may be conserved, to provide a program for the conservation of such endangered species and threatened species, and to take such steps as may be appropriate to achieve the purposes of the treaties and conventions set forth in [subsection a . . . ].

(c) POLICY.—It is further declared to be the policy of Congress that all Federal departments and agencies shall seek to conserve endangered species and threatened species and shall utilize their authorities in furtherance of the purposes of this Act.

*Source:* Public Law 93–205, *United States Statutes at Large*, Vol. 87 (Washington, D.C.: Government Printing Office, 1973), 93rd Cong., 2nd sess., December 28, 1973, pp. 884–85.

## DOCUMENT 120: Ernest Callenbach's *Ecotopia* (1975)

In Ernest Callenbach's novel *Ecotopia*, a skeptical American reporter describes the application of utopian environmental ideas in a secessionist Northwest of the future. It brings to mind the many utopian environmental communes that were formed in the 1960s and 1970s, most of which did not survive into the 1990s. It also reflects the interest in organic farming and sustainable development that took hold during that era.

Wood is a major factor in the topsy-turvy Ecotopian economy, as the source not only of lumber and paper but also of some of the remarkable plastics that Ecotopian scientists have developed. Ecotopians in the city and country alike take a deep and lasting interest in wood. They love to smell it, feel it, carve it, polish it. Inquiries about why they persist in using such an outdated material (which of course has been entirely obsoleted by aluminum and plastics in the United States) receive heated replies. To ensure a stable long-term supply of wood, the Ecotopians early reforested enormous areas that had been cut over by logging companies before Independence. They also planted trees on many hundreds of thousands of acres that had once been cleared for orchards or fields, but had gone wild or lay unused because of the exodus of people from the country into the cities.

I have now been able to visit one of the forest camps that carry out lumbering and tree-planting, and have observed how far the Ecotopians carry their love of trees. They do no clear-cutting at all, and their forests contain not only mixed ages but also mixed species of trees. They argue that the costs of mature-tree cutting are actually less per board foot than clear-cutting—but that even if they weren't it would still be desirable because of less insect damage, less erosion, and more rapid growth of timber. But such arguments are probably only a sophisticated rationale for attitudes that can almost be called tree worship.

\* \* \*

Our economists would surely find the Ecotopian lumber industry a labyrinth of contradictions. An observer like myself can come only to general conclusions. Certainly Ecotopians regard trees as being alive in almost a human sense—once I saw a quite ordinary-looking young man, not visibly drugged, lean against a large oak and mutter "Brother Tree!" And equally certainly, lumber in Ecotopia is cheap and plentiful, whatever the unorthodox means used to produce it. Wood therefore takes the place that aluminum, bituminous facings, and many other modern materials occupy with us.

An important by-product of the Ecotopian forestry policies is that extensive areas, too steep or rugged to be lumbered without causing erosion, have been assigned wilderness status. There all logging and fire roads have been eradicated. Such areas are now used only for camping and as wildlife preserves, and a higher risk of forest fire is apparently accepted.

*Source*: Ernest Callenbach, *Ecotopia* (Berkeley, CA: Banyan Tree Books, 1975), pp. 55, 57–58.

## DOCUMENT 121: Edward Abbey's *The Monkey Wrench Gang* (1975)

The more radical environmental groups, inspired by Thoreau and borrowing their tactics from the civil rights and anti–Vietnam War protesters, engaged in civil disobedience and even sabotage to bring their environmental message to the government and the public and to urge industry to stop its environmentally destructive behavior.

Edward Abbey's fact-based novel *The Monkey Wrench Gang*, which describes the acts of sabotage committed by a group of eco-warriors, acquired a cult following and has served as a handbook for some of the most radical activists. In October 1998, a group of activists, in an effort to stem the expansion of the ski resort at Vail Mountain in Colorado, set fire to buildings and lifts, causing an estimated $12 million in damages. Earth First!, whose members have chained themselves to trees to prevent logging activity, condoned the sabotage.[4]

In this selection from Abbey's novel, Glen Canyon Dam's intrusive presence in the landscape offers sufficient justification for one of the gang members to make plans to blow it up. In 1997 some less radical Americans, in a House Resources Committee hearing on Lake Powell, actually proposed that Congress take action to restore Glen Canyon by draining the lake.[5]

They passed the Wahweap Marina turnoff. Miles away down the long slope of sand, slickrock, blackbrush, Indian ricegrass and prickly pear they could see a cluster of buildings, a house-trailer compound, roads, docks and clusters of boats on the blue bay of the lake. Lake Powell, Jewel of the Colorado, 180 miles of reservoir walled in by bare rock.

The blue death, Smith called it. Like Hayduke his heart was full of a healthy hatred. Because Smith remembered something different. He remembered the golden river flowing to the sea. He remembered canyons called Hidden Passage and Salvation and Last Chance and Forbidden and Twilight and many many more, some that never had a name. He remembered the strange great amphitheaters called Music Temple and Cathedral in the Desert. All these things now lay beneath the dead water of the reservoir, slowly disappearing under layers of descending silt. How could he forget? He had seen too much.

Now they came, amidst an increasing flow of automobile and truck traffic, to the bridge and Glen Canyon Dam. Smith parked his truck in front of the Senator Carl Hayden Memorial Building. He and his friend got out and walked along the rail to the center of the bridge.

Seven hundred feet below streamed what was left of the original river, the greenish waters that emerged, through intake, pen-stock, turbine and tunnel, from the powerhouse at the base of the dam. Thickets of power cables, each strand as big around as a man's arm, climbed the canyon walls on steel towers, merged in a maze of transformer stations, then splayed out toward the south and west—toward Albuquerque, Babylon, Phoenix, Gomorrah, Los Angeles, Sodom, Las Vegas, Nineveh, Tucson, the cities of the plain.

Upriver from the bridge stood the dam, a glissade of featureless concrete sweeping seven hundred feet down in a concave facade from the dam's rim to the green-grass lawn on the roof of the power plant below.

They stared at it. The dam demanded attention. It was a magnificent mass of cement. Vital statistics: 792,000 tons of concrete aggregate; cost $750 million and the lives of sixteen (16) workmen. Four years in the making, prime contractor Morrison-Knudsen, Inc., sponsored by U.S. Bureau of Reclamation, courtesy U.S. taxpayers.

"It's too big," she said.

"That's right, honey," he said. "And that's why."

"You can't."

"There's a way."

*Source*: Edward Abbey, *The Monkey Wrench Gang* (Philadelphia: Lippincott, 1975), pp. 36–37.

---

## DOCUMENT 122: Greenpeace on the Need for Radical Action (1976)

Greenpeace's "Declaration of Interdependence" explains why the organization—which has employed innovative but always nonviolent tactics to effect environmental changes—believed radical activism was necessary. Today Greenpeace, like many other once-radical groups, maintains an office in Washington, D.C., and works closely with the Washington establishment.

We have arrived at a place in history where decisive action must be taken to avoid a general environmental disaster. With nuclear reactors proliferating and over 900 species on the endangered list, there can be no further delay or our children will be denied their future.

The Greenpeace Foundation hopes to stimulate practical, intelligent actions to stem the tide of planetary destruction. We are "rainbow people" representing every race, every nation, every living creature. We are patriots, not of any one nation, state or military alliance, but of the entire earth.

It must be understood that the innocent word "ecology" contains a concept that is as revolutionary as anything since the Copernican breakthrough, when it was discovered that the earth was not the center of the entire universe. Through ecology, science has embarked on a quest for the great systems of order that underlie the complex flow of life on our planet. This quest has taken us far beyond the realm of traditional scientific thought. Like religion, ecology seeks to answer the infinite mysteries of life itself. Harnessing the tools of logic, deduction, analysis, and empiricism, ecology may prove to be the first true science-religion.

As suddenly as Copernicus taught us that the earth was not the center of the universe, ecology teaches us that mankind is not the center of life on this planet. Each species has its function in the scheme of life. Each has a role, however obscure that role may be.

Ecology has taught us that the entire earth is part of our "body" and that we must learn to respect it as much as we respect ourselves. As we love ourselves, we must also love all forms of life in the planetary system—the whales, the seals, the forests and the seas. The tremendous beauty of ecological thought is that it shows us a pathway back to an understanding of the natural world—an understanding that is imperative if we are to avoid a total collapse of the global ecosystem.

Ecology has provided us with many insights. These may be grouped into three basic "Laws of Ecology" which hold true for all forms of life—fish, plants, insects, plankton, whales, and man. These laws may be stated as follows:

**The First Law of Ecology** states that all forms of life are interdependent. The prey is as dependent on the predator for the control of its population as the predator is on the prey for a supply of food. . . .

**The Second Law of Ecology** states that the stability (unity, security, harmony, togetherness) of ecosystems is dependent on their diversity (complexity). An ecosystem that contains 100 different species is more stable than an ecosystem that has only three species. Thus the complex tropical rain-forest is more stable than the fragile arctic tundra. . . .

**The Third Law of Ecology** states that all resources (food, water, air, minerals, energy) are finite and there are limits to the growth of all living systems. These limits are finally dictated by the finite size of the earth and the finite input of energy from the sun. . . .

If we ignore the logical implications of these "Laws of Ecology" we will continue to be guilty of crimes against the earth. We will not be judged by men for these crimes, but with a justice meeted [*sic*] out by the earth itself. The destruction of the earth will lead, inevitably, to the destruction of ourselves.

So let us work together to put an end to the destruction of the earth by the forces of human greed and ignorance. Through an understanding of the principles of ecology we must find new directions for the evolution of human values and human institutions. *Short-term economics* must be replaced with actions based on the need for conservation and preservation of the entire global ecosystem. We must learn to live in harmony, not only with our fellow man, but with all the beautiful creatures on this planet.

*Source*: Greenpeace, "Declaration of Interdependence," in Peter C. List, ed., *Radical Environmentalism: Philosophy and Tactics* (Belmont, CA: Wadsworth, 1993), pp. 134–35.

# Part VII

---

# Confronting Economic and Social Realities, 1980–2000

During the 1980s, the Reagan administration, because of its pro-business stance, attempted to eviscerate the Environmental Protection Agency (EPA), by drastically cutting its funding, and to take the teeth out of much of the environmental legislation passed in the 1970s. Ronald Reagan brought into the government a number of antienvironmentalists, including James Watt as secretary of the interior and Anne Gorsuch (Buford) as head of the EPA. Vehemently opposed to the regulation of business, they encouraged groups in industries such as ranching, mining, and logging to advocate the "wise use" of resources and to focus on the need to develop and use resources. In 1981 Watt stated, "I believe some parts of the country are so beautiful they need preserving. But I also believe we need dams and resource development. We will mine more, drill more, cut more timber to use our resources rather than simply to keep them locked up."[1]

The Reagan administration's probusiness stance reflected the nation's concern about its ability to remain a strong contender in the increasingly competitive global marketplace. For more than thirty-five years, Germany and Japan, which had been demilitarized following World War II, were busy building modern factories to replace their war-damaged industrial infrastructures while the United States was still spending large sums to maintain its military power. In the northeastern United States and in the Midwest, factory towns were abandoned and inner cities decayed as factories moved south or overseas in search of cheap labor and freedom from the constraints of environmental regulation.

It was not until the disintegration of the Soviet Union at the end of the 1980s that the United States was willing to turn away from its arms build-up and focus on technological innovation. However, by the 1980s newly industrialized countries in Southeast Asia and the Pacific Rim, including Singapore, Malaysia, and South Korea, and in Latin America, including Brazil and Mexico, had joined Germany and Japan in competing with the United States, and the nation's blue-collar jobs continued to disappear into foreign factories, some owned by American corporations or their subsidiaries and others by entrepreneurs in the newly industrialized countries.

After World War II, many farm jobs also disappeared as farmers sold off their land to real estate developers, and as small farmers found it more and more difficult to compete with giant agribusinesses. Slowly, an increasing portion of the country's food supply was imported, and in 1996, 16.4 percent of the produce consumed by Americans was imported.[2] (From outbreaks of food poisoning, it became evident that some of this imported produce did not meet American food safety standards.) During the same period, however, the service and high-tech sectors of the American economy were expanding and creating new jobs.

Despite the decrease in manufacturing and farm jobs, the possibility of employment continues to attract tens of thousands of foreign immigrants to the United States every year. In recent years, while the population of most countries in the industrialized world and in eastern Europe has stabilized or is even declining, the U.S. population has continued to increase as a result of immigration pressures from Latin America and Asia, where populations are still burgeoning. Although the rate of U.S. population growth has declined during the past few years, the actual numbers of people continue to rise rapidly. Today there are 268 million people in the United States—89 million more than in 1960 and more than double the nation's population in 1940. The proportion of elderly and very elderly people in the U.S. population is also increasing as the nation's already high life-expectancy rate (71.8 years for men and 78.8 years for women in 1990) continues to rise.

## ENVIRONMENT AND THE AMERICAN LIFESTYLE

By the 1980s the negative impact of population growth on environmental quality had become evident to most Americans. Although individual rights advocates and businesses adversely affected by specific legislative or judicial actions objected to the imposition of environmental controls, few people questioned the need for some conservation of resources or for clean air and clean water standards. While such arch opponents of stringent environmental controls as Dixy Lee Ray [see Document 136] comprehended that the profligate use of natural resources was unhealthy for the nation, they questioned the extent of the impact on the environment of various human activities—from waste disposal to driving to the use of nuclear energy. How people view the potential risk of a particular environmental hazard is frequently based on how the hazard or its elimination will affect their life: Is there a chance that it could affect them economically, in a positive or a negative way? Is there a chance that it could affect their health or the health of their children? Is there a chance that it will impinge on their freedom—to use their land, to drive, or to smoke, for example?

Some economists, such as Julian Simon [see Document 126], remain convinced that there is no need to panic about the effect of population growth on resource use because human technological innovation will enable us to overcome the expanding need for food and other resources. His vision is supported by biotechnologists who foresee the development of bioengineered plants and animals that will be disease resistant and highly productive. Simon and other "wise use" advocates are committed to the development of America's natural resources for the benefit of its human inhabitants. They believe that the increasingly human-dominated environment allows the nation to set aside only limited quantities of material for future emergencies.

Environmentalists like Lester Brown [see Document 125], however, insist that we must give greater value to nature's services and must adjust our appetites to the functioning of natural processes and the limits of the natural world. Both Edward Wilson and Al Gore propose the development of a new conservation ethic based on a biocentric approach to living on earth [see Documents 132, 141], and Wilson has demanded that people look at themselves as biological as well as cultural beings. While many of the established environmental groups, from the Sierra Club to the Environmental Defense Fund, support this kind of thinking, their goal is to effect change without really upsetting the status quo. Even such once radical groups as Greenpeace have realigned their sights in recent years and are working closely with Washington policymakers and environmental regulatory agencies.

More radical ecologists and environmentalists, though, advocate a vast change in the American lifestyle. They claim that the human race must quickly take account of the needs of the natural world and turn from being ignorant predators into thoughtful custodians of other living things on this planet. Because we are destroying the only environment suitable for human habitation or because the environment we are creating is not one they would want to live in [see Documents 124 and 128, and 140], they believe that radical action must be taken immediately.

## THE COMPLEXITY OF ENVIRONMENTAL ISSUES

Many of the environmental problems facing the United States in the 1990s do not have simple solutions. Resolving them entails finding acceptable balances among individual rights, societal needs, economic priorities, ethical values, and the limits of the natural world. For example, one pressing issue is how to satisfy growing demands for energy in the United States as well as in the emerging economies of the world. Because the quantity of energy used correlates to some extent with how industrially developed a country is, as more countries advance industrially, not only will energy use increase, but so will the accompanying pollution and environmental degradation—unless the world as a whole becomes less dependent on fossil fuels. Compounding this is the disproportionately high per capita use of energy in the United States compared to other industrially advanced nations.

Nuclear energy is one possible solution to the energy problem. However, because of the risks posed by the use of nuclear energy, there has been strenuous opposition to the construction of new nuclear power plants [see Document 130]. Although renewable energy sources and clean fuels exist, their high cost and low efficiency impose barriers to their use [see Document 139]. California, which—because of its air pollution problems and dependence on cars—has some of the strictest auto emissions regulations in the country, has had to roll back plans to attain emission reductions [see Document 138] because automobiles that met those standards proved too expensive or unsuitable for consumers. American automakers may be willing to put effort into the development of fuel-efficient cars [see Document 137], but they will put a greater effort into the production of gas-guzzling vehicles—such as sport utility vehicles and pickup trucks—as long as sales of the latter are more profitable and as long as there continues to be a market for them. Furthermore, while auto fuel efficiency and emissions quality have steadily improved, the number of vehicles on American roads has continued to climb (between 1980 and 1995 the total number of cars and trucks in use increased by 53.6 million vehicles), thereby reducing air quality gains.

Another complex problem is that of land use. As the nation's population increases, there is constant demand for land on which to build homes. But the building of homes on former farmland pushes agricultural regions farther and farther from population centers, and the construction of homes in wilderness areas endangers wildlife habitats and shrinks the nation's open space. The use of coastal regions and wetlands areas for home construction not only endangers precarious ecosystems but also poses the issue of the insurability of waterfront property that is subject to shifting water courses and changing land profiles [see Document 127]. Using arid lands for home construction, agriculture, or grazing creates a host of new problems, from the need for an adequate water supply to the unanticipated consequences of destabilizing a fragile landscape. The 1980s and 1990s have seen a growing movement to protect farmland and open space as more and more people in suburban and rural areas realize that development is eating away at their cherished way of life and the places with which they have a bond. [see Document 140]. In November 1998, there were more than 200 different questions on ballots across the nation on issues relating to open space and farmland conservation programs.

A third complex problem is that of waste disposal. The need to dispose of increasing amounts of household wastes (including the ever-increasing trash from packaged goods); a wide and growing range of toxic chemicals employed in industry, agriculture, and the home; and accumulating radioactive wastes pose problems of where to put the wastes, how to dispose safely of dangerous wastes, and how to deal with pollution from improperly disposed of toxic materials [see Documents 123 and 135].

In the 1980s a great deal of attention was given to the development of recycling programs. Certainly recycling helps reduce the space needed for waste disposal, and many people consider recycling a useful tool for conserving natural resources and see trash as a valuable source of energy, but others question the economic and physical efficiency of recycling.

Unfortunately, in the United States, as elsewhere in the world, the response to complex environmental issues is frequently driven by short-term economic and social pressures. Those who have money and power insist on having clean water, clean air, and unpolluted land in their backyards. It is the poor and minorities who are most likely to live in environmentally degraded neighborhoods [see Document 135].

As people have introduced new technologies and products and have altered the landscape around them, they have also produced many unforeseen changes in their environment. During the past few decades, there have been numerous efforts to undo many of these alterations, including some that were intentional and others that were accidental. In Marin County in northern California, whole forests of eucalyptus trees (an invasive introduced species) are being cut down and replaced with native oaks. In south Florida efforts are underway to undo some of the channelization of the waters flowing into and out of the Kissimmee River, Lake Okeechobee, and the Everglades system that had been instituted over the past century [see Document 131]. Recently there has even been serious talk of undoing such monumental projects as Glen Canyon Dam on the Colorado River in northern Arizona. However, over time, the presence of foreign species of plants and animals or of man-made structures that interfere with water courses may themselves have effected changes that will impact the restoration projects.

While many technological advances and scientific innovations—

from great dams to wonder drugs—have provided immediate benefits to large numbers of people, we are beginning to discover that some of these "improvements" have merely shifted the nature of the environmental problems with which we have to cope [see Document 144]. The biotechnology revolution offers the possibility of creating ever more productive food plants and animals and eliminating genetic diseases, but because these new organisms and gene changes involve the manipulation of nature at the most fundamental level, the revolution is sure to be accompanied by a host of unforeseen environmental problems [see Document 146].

## CONFRONTING THE GLOBAL ENVIRONMENTAL EFFECTS OF LOCAL HUMAN ACTIVITY

The task of balancing competing priorities is complicated by the global environmental effects of many local and national activities. Population expansion, energy use, hunting and fishing of migratory species, wetlands use, waste disposal, and industrial pollution can have international repercussions.

The recognition that environmental degradation is a global concern has resulted in a multiplicity of international declarations, protocols, and treaties on such issues as acid rain, climate change, marine pollution, the use of marine resources, and ozone protection [see Documents 129, 133, 134, and 142]. But as Jurgen Schmandt, Hilliard Roderick, and Andrew Morriss point out, such agreements are meaningless unless all the parties to an agreement impose national rulings to enforce the agreement [Document 133]. Obviously, nations will not undertake such action if the agreement appears to be detrimental to their own economic or political interests.

## ENVIRONMENTAL CHALLENGES FOR THE TWENTY-FIRST CENTURY

As the millennium approaches, apocalyptic voices predicting uncontrollable global epidemics, the destruction of the earth's atmosphere, global warming [see Document 142], the dwindling of adequate and safe water supplies, and the disappearance of fish [see Document 143] are heard with increasing frequency. For the past century environmental regulations instituted by the United States and the international community have resulted in improvements in the quality of life of large numbers of people. However, we are slowly exhausting the possibilities of simple, financially acceptable environmental controls.

Despite the many efforts to rescind green legislation and to limit the reach of environmental regulation, it appears that most Americans are now convinced that the rights of the populace to a clean, safe environment may outweigh the rights of individuals to use their private property in any way they choose. In the twenty-first century the United States will be forced to reexamine its priorities—it may have to give up the idea of continuous economic growth, for example—if it wants to maintain both economic and societal well-being [see Document 145]. As we enter "the century of the environment," the United States will have to recognize that how it defines its relationship to the natural world will affect not only its ability to be a sustainable society but also its ability to play an influential role on a planet that can sustain human life.

---

## DOCUMENT 123: Comprehensive Environmental Response, Compensation, and Liability Act (1980)

In 1978 the community of Love Canal near Niagara Falls, New York, had to be evacuated because toxic wastes dumped in the region had created a serious health hazard. Partly in response to this disaster, Congress passed the Comprehensive Environmental Response, Compensation, and Liability Act, better known as the Superfund Act. During the nearly twenty years that this law has been in effect, attempts have been made to clean up a few notorious waste sites, but because of the difficulty and cost of toxic waste removal, both industry and government have frequently sought ways to circumvent the law.

SEC. 104. (a) (1) Whenever (A) any hazardous substance is released or there is a substantial threat of such a release into the environment, or (B) there is a release or substantial threat of release into the environment of any pollutant or contaminant which may present an imminent and substantial danger to the public health or welfare, the President is authorized to act, consistent with the national contingency plan, to remove or arrange for the removal of, and provide for remedial action relating to such hazardous substance, pollutant or contaminant at any time (including its removal from any contaminated natural resource), or take any other response measure consistent with the national contingency plan which the President deems necessary to protect the public health or welfare or the environment, unless the President determines that such removal and remedial action will be done properly by the owner or operator of the vessel, or facility from which the release or threat of release emanates, or by any other responsible party.

(2) For the purposes of this section, "pollutant or contaminant" shall

include, but not be limited to, any element, substance, compound, or mixture, including disease-causing agents, which after release into the environment and upon exposure, ingestion, inhalation, or assimilation into any organism, either directly from the environment or indirectly by ingestion through food chains, will or may reasonably be anticipated to cause death, disease, behavioral abnormalities, cancer, genetic mutation, physiological malfunctions (including malfunctions in reproduction) or physical deformations, in such organisms or their offspring. The term does not include petroleum, including crude oil and any faction thereof which is not otherwise specifically listed or designated as hazardous substances under section 101 . . . of this title, nor does it include natural gas, liquefied natural gas, or synthetic gas of pipeline quality. . . .

SEC. 105. Within one hundred and eighty days after the enactment of this Act, the President shall, after notice and opportunity for public comments, revise and republish the national contingency plan for the removal of oil and hazardous substances, originally prepared and published pursuant to section 311 of the Federal Water Pollution Control Act, to reflect and effectuate the responsibilities and powers created by this Act. . . . Such revision shall include a section of the plan to be known as the national hazardous substance response plan which shall establish procedures and standards for responding to releases of hazardous substances, pollutants, and contaminants, which shall include at a minimum:

(1) methods for discovering and investigating facilities at which hazardous substances have been disposed of or otherwise come to be located;

(2) methods for evaluating, including analyses of relative cost, and remedying any releases or threats of releases from facilities which pose substantial danger to the public health or the environment;

(3) methods and criteria for determining the appropriate extent of removal, remedy, and other measures authorized by this Act;

(4) appropriate roles and responsibilities for the Federal, State, and local governments, and for interstate and nongovernmental entities in effectuating the plan;

(5) provision for identification, procurement, maintenance, and storage of response equipment and supplies;

(6) a method for and assignment of responsibility for reporting the existence of such facilities which may be located on federally owned or controlled properties and any releases of hazardous substances from such facilities;

(7) means of assuring that remedial action measures are cost-effective over the period of potential exposure to the hazardous substances or contaminated materials;

(8) (A) criteria for determining priorities among releases or threatened releases throughout the United States for the purpose of taking remedial action, and, to the extent practicable taking into account the potential urgency of such action,

for the purpose of taking removal action. Criteria and priorities under this paragraph shall be based upon relative risk or danger to public health or welfare or the environment, in the judgment of the President, taking into account to the extent possible the population at risk, the hazard potential of the hazardous substances at such facilities, the potential for contamination of drinking water supplies, the potential for direct human contact, the potential for destruction of sensitive ecosystems, State preparedness to assume State costs and responsibilities, and other appropriate factors.

(B) based upon the criteria set forth in subparagraph (A) of this paragraph, the President shall list as part of the plan national priorities among the known releases or threatened release throughout the United States and shall revise the list no less often than annually. Within one year after the date of enactment of this Act, and annually thereafter, each State shall establish and submit for consideration by the President priorities for remedial action among known releases and potential releases in that State. . . .

(9) specific roles for private organizations and entities in preparation for response and in responding to releases of hazardous substances.

*Source*: Public Law 96–510, *United States Statutes at Large*, Vol. 94, Part 3 (Washington, D.C.: Government Printing Office, 1981), 96th Cong., December 11, 1980, pp. 2774–75, 2779–80.

---

## DOCUMENT 124: Mark Sagoff on the Public Interest (1981)

The political philosopher Mark Sagoff objects to economic efficiency being the sole criterion for determining government policy because it does not distinguish between public and private interests. Economic efficiency, he notes, takes into account only private, self-serving interests (typically to be satisfied in markets), and ignores morality, justice, and spirituality, which are fundamental public values that should be determined through a deliberative political process.

Writing in the June 1997 issue of the *Atlantic Monthly*, he posited that "the world has the wealth and the resources to provide everyone the opportunity to live a decent life. We consume too much when market relationships displace the bonds of community, compassion, culture, and place. We consume too much when consumption becomes an end in itself and makes us lose affection and reverence for the natural world."[3]

Many economists take the view that environmental problems are economic problems. They believe that market failure causes these problems: private and social costs diverge; profit-maximizing decisions, therefore, are socially inefficient. Economists would correct this market failure by requiring private decision makers to internalize externalities, that is, to make the price of goods reflect all the economic and social costs of producing them, including the pollution costs. When this is done, they argue, pollution will be controlled, endangered species will be saved, and pristine areas will be preserved, but only to the extent that the benefits exceed the costs....

Although this economic approach purports to allow us to choose the best among available policies, in fact it makes economic efficiency our only goal. Economic efficiency has traditionally been understood to require the maximum satisfaction of the preferences that markets reveal. These are typically self-regarding or self-interested preferences, that is, preferences that reflect a person's idea of his or her individual welfare. Preferences of this sort may be contrasted with preferences that express what the individual believes is in the public interest or in the interest of a group or community to which he or she belongs. Political activity is supposed, in theory at least, to provide a vehicle for airing, criticizing, and settling upon interests or opinions of this group-regarding kind.

The search for economic efficiency might take us to the best public policies if we were a nation of individualists competing each for his or her own welfare with no regard for or conception of the collective good. Then an efficient market might lead us to satisfy as well as possible all of our interests. In such a situation, government might best be conceived as a prophylactic on markets, and public policies might be considered irrational if they could not be construed as reasonable responses to market failures. But we are not simply a group of consumers, nor are we bent on satisfying only self-regarding preferences. Many of us advocate ideas and have a vision of what we should do or be like as a nation. And we would sacrifice some of our private interests for those public ends.

...Why should we believe that the right policy goal is the one that satisfies only the self-interested preferences of consumers? Why should we not take into account the community-regarding values that individuals seek through the political process as well?

* * *

Anyone who believes that government ought to be primarily interested in correcting market failure must find puzzling much of our environmental legislation. Environmentalist groups, not famous for their economic "common sense," successfully backed much of this legislation in the 1970s. It is not surprising, therefore, that environmental protection goes beyond the mere correction of market deficiencies. Congress designed the Clean Air and Clean Water Acts to improve the quality of our air and water. It passed the Endangered Species Act to protect threatened species, even if the economic costs of protection outweigh the benefits. Similarly, the Occupational Safety and Health Act seeks to make the workplace safe and healthful, a goal that is not always consistent with market efficiency.

\* \* \*

Although economic approaches to public policy may purport to weigh both consumer and citizen values, we may, as citizens, believe that certain public values or collective goals (*e.g.*, that an innocent person not be convicted) supersede the values that we pursue as self-seeking individuals (*e.g.*, security from crime). Moreover, we might decide to sacrifice economic optimality for cleaner air and water. Once legislatures, responding to political pressure, have made this choice, is it defensible for economists to insist that our policymaking process include the very consumer values that we have decided to sacrifice?

\* \* \*

When an environmentalist argues that we ought to preserve wilderness areas because of their cultural importance and symbolic meaning, he or she states a *conviction* and not a *desire*. When an economist asserts that we ought to attain efficient levels of pollution, he or she, too, states a belief. Both beliefs are supported by arguments, not money.

\* \* \*

What many economists do not understand is that efficiency is one value among many and is not a meta-value that comprehends all others. Economists as a rule do recognize one other value, namely, justice or equality, and they speak, therefore, of a "trade-off" between efficiency and equality. They do not speak, as they should, however, about the trade-off between efficiency and our aesthetic and moral values. What about the trade-off between efficiency and dignity, efficiency and self-respect, efficiency and the magnificence of our natural heritage, efficiency and the quality of life? These are the trade-offs that are important in setting environmental policy.

*Source*: Mark Sagoff, "Economic Theory and Environmental Law," *Michigan Law Review* 79 (1981): 1393–95, 1399, 1416, 1419.

## DOCUMENT 125: Lester R. Brown on Building a Sustainable Society (1981)

In 1974 Lester Brown, who holds a degree in agriculture and at one time worked for the U.S. Department of Agriculture, founded the Worldwatch Institute, an organization devoted to worldwide environmental issues relating to sustainability and global interdependency. The institute's influential annual report on the environment, *State of the World*, is a sourcebook for corporate leaders and government policymakers around the world.

The concept of sustainable development is an outgrowth of ideas developed by conservationist-minded forest and land managers such as Gifford Pinchot and John Wesley Powell, who, at the end of the nineteenth century, called for the wise use of America's resources. Interest in sustainable development gained momentum in the 1970s, but the focus of the new breed of conservationists who supported this concept was on international economic, energy, and resource policies and on activities that would make sustainability possible.

The international community's commitment to sustainable development was made evident in the Stockholm Declaration [see Document 118] and was clearly enunciated in both the 1987 Brundtland Report (the United Nations World Commission on Environment and Development's plan for nations to find areas of agreement on environmental issues that involve the interplay of environmental and economic factors, which extended the concept of sustainable development to the entire globe), and the 1992 United Nations Conference on Environment and Development (UNCED) "Programme of Action for Sustainable Development" (also known as the Rio Declaration).[4]

A sustainable society will differ from the one we now know in several respects. Population size will more or less be stationary, energy will be used far more efficiently, and the economy will be fueled largely with renewable sources of energy. As a result, people and industrial activity will be more widely dispersed, far less concentrated in urban agglomerations than they are in a petroleum-fueled society.

The transition to renewable energy will endow the global economy with a permanence that coal and oil-based societies lack. More than that, it could lead us out of an inequitable, inherently unstable international energy regime since, unlike coal and oil, solar energy is diffuse, available in many forms, and accessible to all countries.

As the switch from fossil energy to solar energy progresses, the geographic distribution of economic activity is destined to change, conforming to the location of the new energy sources. The transition to a sustainable society promises to reshape diets, the distribution of population, and modes of transportation. It seems likely to alter rural-urban relationships within countries and the competitive position of national economies in the world market. Then too, a sustainable society will require labor force skills markedly different from those of the current oil-based economy.

\* \* \*

Before us now is the opportunity to adjust our values according to our changing perceptions of our world and our place in it. Of necessity, the path to sustainability will be littered with cast-off values. Materialism, planned obsolescence, and a desire for large families will not survive the transition. But they will not leave a void. Frugality, a desire for a harmonious relationship with nature, and other values compatible with a sustainable society will take their place.

*Source*: Lester R. Brown, *Building a Sustainable Society* (New York: Norton, 1981), pp. 247–48, 350.

## DOCUMENT 126: Julian L. Simon on Population Growth (1981)

Many people disagree with the apocalyptic environmentalists such as Paul Ehrlich [see Document 106] and the radical environmentalists such as Arne Naess [see Document 128] who say that we must take immediate action to constrain the human impact on the environment. One of the most strident opponents of their ideas is Julian Simon, a professor of economics and business administration at the University of Illinois at the time that he wrote *The Ultimate Resource*, from which this selection is taken, as a popularization of his earlier book about population growth. Simon later moved to the Heritage Foundation, a conservative think-tank in Washington, D.C. Some of Simon's ideas have roots in the writings of John Locke [see Document 14], who contended that how well a person used his land was more important than how much property he possessed.

*Food*. Contrary to popular impression, the per capita food situation has been improving for the three decades since World War II, the only decades for which we have acceptable data. We also know that famine has progressively diminished for at least the past century. And there is strong reason to believe that human nutrition will continue to improve into the indefinite future, even with continued population growth.

*Land.* Agricultural land is not a fixed resource, as Malthus [see Document 26] and many since Malthus have thought. Rather, the amount of agricultural land has been, and still is, increasing substantially, and it is likely to continue to increase where needed. Paradoxically, in the countries that are best supplied with food, such as the U.S., the quantity of land under cultivation has been decreasing because it is more economical to raise larger yields on less land than to increase the total amount of farmland. For this reason, among others, land for recreation and for wildlife has been increasing rapidly in the U.S. All this may be hard to believe, but solid data substantiate these statements beyond a doubt.

*Natural resources.* Hold your hat—our supplies of natural resources are not finite in any economic sense. Nor does past experience give reason to expect natural resources to become more scarce. Rather, if the past is any guide, natural resources will progressively become less scarce, and less costly, and will constitute a smaller proportion of our expenses in future years. And population growth is likely to have a long-run *beneficial* impact on the natural-resource situation.

*Energy.* Grab your hat again—the long-run future of our energy supply is at least as bright as that of other natural resources, though political maneuvering can temporarily boost prices from time to time. Finiteness is no problem here either. And the long-run impact of additional people is likely to speed the development of a cheap energy supply that is almost inexhaustible.

*Pollution.* This set of issues is as complicated as you wish to make it. But even many ecologists, as well as the bulk of economists, agree that population growth is not the villain in the creation and reduction of pollution. And the key trend is that life expectancy, which is the best over-all index of pollution level, has improved markedly as the world's population has grown.

* * *

There is no physical or economic reason why human resourcefulness and enterprise cannot forever continue to respond to impending shortages and existing problems with new expedients that, after an adjustment period, leave us better off than before the problem arose. Adding more people will cause us more such problems, but at the same time there will be more people to solve these problems and leave us with the bonus of lower costs and less scarcity in the long run. The bonus applies to such desirable resources as better health, more wilderness, cheaper energy, and a cleaner environment.

*Source*: Julian L. Simon, *The Ultimate Resource* (Princeton, NJ: Princeton University Press, 1981), pp. 5–6, 345–46.

## DOCUMENT 127: Coastal Barrier Resources Act (1981)

Much of the energy of the environmental movement in the 1960s was concentrated on promoting legislation to protect oceanfront beaches and the salt marshes behind them. In 1972, the same year that the Clean Water Act was passed, Congress approved the Coastal Zone Management Act, a measure designed to convince states to develop broad management plans for their coastal regions and help them pay for the planning process. However, the bill did not directly address any major problems associated with the development of the barrier islands, including the risks to wildlife or the enormous federal expenses that would stem from development. These issues were left for the Coastal Barrier Resources Act, passed in 1981.

The bill clearly affected how private property along the coast would be used. Although it did not restrict land use, it did remove some federal support for the development of coastal lands, and therefore real-estate interests opposed it. Environmentalists favored the Coastal Barrier Resources Act because it promoted the conservation of island and salt marsh habitats; others supported it because they believed it would reduce catastrophic property damage and loss of life due to hurricanes; and still others saw the bill as a way to reduce federal spending by eliminating federal insurance protection for residences and businesses built in high-risk areas (where, it was estimated, payments for damages in 1980 exceeded premium payments by three to one).

At the hearings preceding the vote on the bill, two of the key speakers were James W. Pulliam, Jr., deputy associate director, National Wildlife Refuge System, U.S. Fish and Wildlife Service, who favored the bill, and Lawrence Young, representing the National Association of Realtors, who opposed it. Their statements highlight the fundamental conflict between wildlife managers and conservationists and land developers.

### A. The Bill

(a) FINDINGS.—The Congress finds that—

(1) coastal barriers along the Atlantic and gulf coasts of the United States and the adjacent wetlands, marshes, estuaries, inlets, and near-shore waters provide—

(A) habitats for migratory birds and other wildlife; and

(B) habitats which are essential spawning, nursery, nesting, and feeding areas for commercially and recreationally important species of finfish and shellfish, as well as other aquatic organisms such as sea turtles;

(2) coastal barriers contain resources of extraordinary scenic, scientific, recreational, natural, historic, archeological, cultural, and economic importance, which are being irretrievably damaged and lost due to development on, among, and adjacent to, such barriers;

(3) coastal barriers serve as natural storm protective buffers and are generally unsuitable for development because they are vulnerable to hurricane and other storm damage and because natural shoreline recession and the movement of unstable sediments undermine manmade structures;

(4) certain actions and programs of the Federal Government have subsidized and encouraged development on coastal barriers and the result has been the loss of barrier resources, threats to human life, health, and property, and the recurring expenditure of millions of tax dollars; and

(5) a program of coordinated action by Federal, State, and local governments is critical to the more appropriate use and conservation of coastal barrier resources.

(b) PURPOSE.—The Congress declares that it is the purpose of this Act to minimize the loss of human life, wasteful expenditure of Federal revenues, and damage to fish and wildlife and other resources associated with the coastal barriers along the Atlantic and gulf coasts by establishing a Coastal Barrier Resources System, by restricting future Federal expenditures and financial assistance which have the effect of encouraging development of coastal barriers, and by considering the means and measures by which the long-term conservation of these coastal barrier resources may be achieved.

## B. Statement of James W. Pulliam, Jr.

The great concentrations and diversity of fish and wildlife associated with the relatively limited area of coastal barriers can, in part, be attributed to the role of these landforms as the terrestrial buffer between protected estuaries and lagoons and the more turbulent nearshore ocean waters. These waters, including estuarine and lagoon marshes, intertidal beaches and tidal flats, are among the most fertile and productive known. . . . Up to 90 percent of all commercially important fin and shellfish caught on the Atlantic and gulf coasts are dependent during some stage of their life cycle on estuarine habitat largely created by coastal barriers. Many of these species, as well as others, constitute the base of a large recreational fishery along the Atlantic and gulf coasts.

. . .

Migratory waterfowl are winter inhabitants of coastal barriers and the waters which lie behind them. Species which rely on these ecosystems include whistling swans, snow and Canada geese, widgeon, gadwall and mallards. . . .

Barriers along the Atlantic and gulf coast are also a key migration route for several raptors. The peregrine falcon and, in particular, the arctic peregrine, use the coastal barriers, feeding primarily on sanderlings, killdeer, flickers, and other medium-sized birds. Other species utilizing these landforms include the bald eagle, merlin, osprey, kestrel, and marsh hawk.

. . .

Over 20 vertebrate species associated with coastal barrier islands have been listed pursuant to the Endangered Species Act of 1973 [see Document 119]. These include such endangered birds as the whooping crane, bald eagle, eastern brown pelican, as well as other wildlife species like the manatee, American crocodile, and the green loggerhead, Ridley, and hawksbill sea turtles.

### C. Statement of Lawrence Young

[T]he position of the National Association of Realtors is that it is opposed to the enactment of the Coastal Barrier Resources Act. . . . This legislation will deny reasonable use of private property and is a de facto Federal land-use bill masquerading as a fiscally conservative measure. The chief aim of this bill is to halt coastal development and Congress should address this legislation on that basis.

*Source: Barrier Islands: Hearings Before the Subcommittee on Fisheries and Wildlife Conservation and the Environment and the Subcommittee on Oceanography of the Committee on Merchant Marine and Fisheries, H.R. 3252, 97th Cong., 1st sess., April 27, 1981, June 23, 1981, June 22, 1982 (Washington, D.C.: Government Printing Office, n.d.), Serial no. 97–37, pp. 9–10; Statements: pp. 24–26, 346.*

## DOCUMENT 128: Arne Naess on Deep Ecology (1982, 1984)

The deep ecologist Arne Naess, a Norwegian philosopher who has taught in the United States, has developed a radical, very personal approach to environmental issues. In his 1973 essay "The Shallow and the Deep, Long-Range Ecology Movements," Naess dismisses the "fight against pollution and resource depletion" as a "Shallow Ecology movement" whose "central objective" is "the health and affluence of people in the developed countries."[5] He proposes that people who are seriously concerned about the environment should reconsider some of their basic assumptions and values. Naess's ideas were little known in the United States until the 1980s, when they were promoted and developed by the philosopher George Sessions and the sociologist Bill Devall. His basic principles of deep ecology—the deep ecology platform—were refined on a camping trip to Death Valley, California, with Sessions in April 1984.

### A. Arne Naess Explains Deep Ecology, 1982

The essence of deep ecology is to ask deeper questions. The adjective "deep" stresses that we ask why and how, where others do not. For instance, ecology as a science does not ask what kind of a society would be the best for maintaining a particular ecosystem—that is considered a question for value theory, for politics, for ethics. As long as ecologists keep narrowly to their science, they do not ask such questions. What we need today is a tremendous expansion of ecological thinking in what I call ecosophy. *Sophy* comes from the Greek term *sophia*, "wisdom," which relates to ethics, norms, rules, and practice. Ecosophy, or deep ecology, then, involves a shift from science to wisdom.

For example, we need to ask questions like, Why do we think that economic growth and high levels of consumption are so important? The conventional answer would be to point to the economic consequences of not having economic growth. But in deep ecology, we ask whether the present society fulfills basic human needs like love and security and access to nature, and, in so doing, we question our society's underlying assumptions. We ask which society, which education, which form of religion, is beneficial for all life on the planet as a whole, and then we ask further what we need to do in order to make the necessary changes. We are not limited to a scientific approach; we have an obligation to verbalize a total view.

### B. The Deep Ecology Platform, 1984

1. The flourishing of human and non-human living beings has value in itself. The value of non-human beings is independent of their usefulness to humans.

2. Richness of kinds of living beings has value in itself.

3. Humans have no right to reduce this richness except to satisfy vital human needs.

4. The flourishing of human life is compatible with a substantial decrease of the human population. The flourishing of non-human life requires such a decrease.

5. Present human interference with the non-human world is excessive, and the situation is worsening.

6. Policies must be changed in view of points (1)–(5). These policies affect basic economic, technological, and ideological structures. The resulting state of human affairs will be greatly different from the present.

7. The appreciation of a high quality of life will supersede that of a high standard of life.

8. Those who accept the foregoing points have an obligation to try to contribute directly to the implementation of necessary changes.

*Source*: **A**. Stephen Bodian, "Simple in Means, Rich in Ends: A Conversation with Arne Naess," *Ten Directions* (Zen Center of Los Angeles), Summer–Fall 1982, in Bill Devall and George Sessions, eds., *Deep Ecology: Living as if Nature Mattered* (Salt Lake City: Peregrine Smith Books, 1985), p. 75. **B**. Arne Naess, "Sustainable Development and Deep Ecology," in J. Ronald Engel and Joan Gibb Engel, eds., *Ethics of Environment and Development: Global Challenge, International Response* (London: John Wiley, 1992), p. 88.

---

## DOCUMENT 129: United Nations Convention on the Law of the Sea (1983)

---

The Law of the Sea is probably the most far reaching of the United Nations environmental agreements. It sets forth the limits of national sovereignty and demands that parties to the agreement abide by international regulation of activity in nonterritorial waters. The days when the high seas could be viewed as a commons in which nations could freely dump garbage and harvest resources are gone [see Document 111B].

Although the United States is a signatory to the convention—meaning that it agrees to adhere fundamentally to the convention—the U.S. Senate has not yet ratified it because of strategic concerns. However, several amendments that address American concerns are being added to the convention, with the hope that the United States will ratify it soon.

Article 2

The sovereignty of a coastal State extends, beyond its land territory and internal waters and, in the case of an archipelagic State, its archipelagic waters, to an adjacent belt of sea, described as the territorial sea.

This sovereignty extends to the air space over the territorial sea as well as to its bed and subsoil.

The sovereignty over the territorial sea is subject to this Convention and to other rules of international law.

## Article 3

Every State has the right to establish the breadth of its territorial sea up to a limit not exceeding 12 nautical miles, measured from baselines determined in accordance with this Convention.

## Article 33

In a zone contiguous to its territorial sea, described as the contiguous zone, the coastal State may exercise the control necessary to:

(a) prevent infringement of its customs, fiscal, immigration or sanitary laws and regulations within its territory or territorial sea;

(b) punish infringement of . . . laws and regulations [set forth in this part of the Convention] committed within its territory or territorial sea.

The contiguous zone may not extend beyond 24 nautical miles from the baselines from which the breadth of the territorial sea is measured.

## Article 55

The exclusive economic zone is an area beyond and adjacent to the territorial sea, subject to the specific legal regime established in this Part, under which the rights and jurisdiction of the coastal State and the rights and freedoms of other States are governed by the relevant provisions of this Convention.

## Article 56

In the exclusive economic zone, the coastal State has

(a) sovereign rights for the purpose of exploring and exploiting, conserving and managing the natural resources, whether living or non-living, of the waters superjacent to the sea-bed and of the sea-bed and its subsoil and with regard to other activities for the economic exploitation and exploration of the zone, such as the production of energy from the water, currents and winds;

(b) jurisdiction as provided for in the relevant provisions of the Convention with regard to:

(i) the establishment and use of artificial islands, installations and structures;

(ii) marine scientific research;

(iii) the protection and preservation of the marine environment. . . .

*Source: The Law of the Sea: Official Text of the United Nations Convention on the Law of the Sea with Annexes and Index* (New York: United Nations, 1983), pp. 3, 11, 18.

## Document 130: Bernard Cohen on Nuclear Energy and Risk Assessment (1983)

Sensible evaluation of environmental issues frequently requires an understanding of complex scientific data in order to asses environmental risk. Alarmists have always been very ready to distort facts by providing one-sided information to the public. In his book *Before It's Too Late*, published four years after the partial meltdown of a nuclear reactor at Three Mile Island in Pennsylvania had stifled further discussion of nuclear expansion in the United States, the physicist Bernard Cohen questions the ability of the public to evaluate the hazardousness of nuclear energy.

How well does the American public understand the hazards of nuclear power? A poll of radiation health scientists shows that 82% of them feel that the public's fear of radiation is "substantially" or "grossly" exaggerated. Another poll shows that 89% of all scientists, and 95% of scientists involved in energy-related fields, favor proceeding with the development of nuclear power; among the public there is only a slight majority in favor.

In a recent study in Oregon, groups of college students and members of the League of Women Voters were asked to rank 30 technologies and activities according to the "present risk of death" they pose to the average American. Both groups ranked nuclear power No. 1, well ahead of motor vehicles, which kill about 50,000 Americans each year, cigarette smoking, which kills 150,000, and eleven others that each kill over 1,000. How many can be expected to die annually from generation of nuclear power including the risk of accidents, radioactive waste, and all of the other dangers we hear so much about? According to estimates developed by government-sponsored research programs, about *ten* per year. If you don't trust "the Establishment," you might trust the leading antinuclear activist organization in the United States, the Union of Concerned Scientists (USC), which estimates an average of 120 deaths per year from nuclear power. In either case it is obvious that nuclear power is perceived to be *thousands of times* more dangerous than it actually is. . . . Clearly the American public is grossly misinformed about the hazards of nuclear power.

\* \* \*

[T]wo of the most serious environmental problems we face are air pollution and acid rain. Air pollution is doing billions of dollars worth of damage each year, spreading filth and ugliness, and destroying a wide variety of property ranging from women's stockings to granite statues. Acid rain is rendering lakes lifeless, damaging the forestry and fishing industries, and creating international tensions between the United States and Canada. Both the air pollution and acid rain problems could be largely eliminated by large-scale use of nuclear power. The tragedy of the misunderstanding deepens.

As we project into the future, the tragedy multiplies. Burning coal, oil, and gas is causing earthshaking climatic changes that could eventually turn our Midwestern grain belt into a desert, and flood out our coastal cities—New York, Miami, New Orleans, Houston, Los Angeles, and a host of others. Nuclear power could prevent this if the misunderstandings about its dangers could be eliminated.

The most important problem for our distant progeny will be a shortage of materials that we now obtain by mining. We are now consuming the world's scarce mineral resources at a voracious rate; indeed our era has been called "the age of mining," because within less than a century there will be very little left to mine—no copper, no tin, no lead, no mercury, no zinc, and so on. In the desperate search for substitutes, the most fruitful source would be plastics and organic chemicals. But these are made from coal, oil, and gas, which we are now simply burning up at a rate of millions of tons each [and] every day. Wouldn't it be much better if we instead burned uranium, which has no other important uses, leaving the coal, oil, and gas for future generations to use as a source of materials they will so sorely need?

*Source*: Bernard L. Cohen, *Before It's Too Late: A Scientist's Case for Nuclear Energy* (New York: Plenum, 1983), pp. 1–2, 4–5.

---

## DOCUMENT 131: Bob Graham on Restoring the Kissimmee River–Lake Okeechobee–Everglades Ecosystem (1983)

For more than a hundred years, south Florida wetlands have been subject to draining, channelization, and other manipulations that put severe stress on the region's entire ecosystem. In August 1983, Florida's governor, Bob Graham, proposed a long-term restoration program to save the Everglades, and in November he issued an executive order to move the program forward. Graham's program, which took into account the needs of the whole south Florida ecosystem, was the kind of wide-ranging program for the area that Marjory Stoneman Douglas had advocated in the 1940s [see Document 85] but that, at the time, failed to gain adequate public support.

The Kissimmee River–Lake Okeechobee–Everglades plan is but one of several environmental restoration projects undertaken around the country to undo carefully planned and frequently government sanctioned manipulations of ecosystems.

## A. "Save Our Everglades" Issue Paper

The Kissimmee River, once gently meandering for 90 miles from Lake Kissimmee to Lake Okeechobee, was channelized in the 1960's by the U.S. Army Corps of Engineers at the request of the State. This one-time paradise of fish and wildlife is now a 48-mile canal, 30 feet deep and 200 feet wide, commonly known as the "Kissimmee Ditch." Channelization of the Kissimmee directly destroyed 40,000 acres of river marsh and allowed drainage of more than 100,000 acres of associated wetlands. A once-serpentine river of pristine quality has become a discharge canal into Lake Okeechobee.

## B. Executive Order

WHEREAS, it is the policy of the State of Florida to protect and manage the unique Central and Southern Florida natural resources of the Kissimmee River–Lake Okeechobee–Everglades ecosystem, in order to enhance existing ecological, recreational, scientific, economic, water supply, and flood control values for present and future Floridians, and

WHEREAS, the future of both the systems of man and nature in Southern Florida depend upon the restoration and enhancement of the functioning Kissimmee River–Lake Okeechobee–Everglades ecosystems, and

WHEREAS, the water resources and ecological health of these systems are extremely vulnerable to development and sensitive to management activities, and

WHEREAS, in their natural condition, these ecosystems perform critical water resource services—and provide natural and free and renewable benefits—in terms of flood control, water treatment, water storage and supply, and aquifer recharge, and

WHEREAS, many species of Florida wildlife, including endangered species, depend upon the Kissimmee River–Lake Okeechobee–Everglades ecosystem for habitat, and

WHEREAS, these ecosystems provide a unique source of natural beauty, wilderness refuge, and recreational enjoyment for millions of residents and visitors, and

WHEREAS, much of the past utilization of lands and waters within these ecosystems has been destructive to ecologically sensitive resources, and

WHEREAS, certain federal, state and regional and local activities, programs and management policies have historically subsidized and encouraged development within these ecosystems resulting in significant destruction of wetlands and other valuable natural resources, and

WHEREAS, various federal, state, regional and local agencies of government are presently involved in a wide range of resource planning and management activities aimed at the protection, restoration and enhancement of the natural values of the Kissimmee River–Lake Okeechobee–Everglades ecosystems.

Now, therefore, I, Bob Graham . . . do hereby promulgate the following Executive Order effective immediately:

The Kissimmee River–Lake Okeechobee–Everglades Coordinating Council (KOECC) is hereby created. . . .

The KOECC is created for the purpose of coordination and promotion of restoration efforts in the Kissimmee River-Lake Okeechobee–Everglades ecosystems. The purpose shall encompass the following objectives for the Kissimmee River–Lake Okeechobee–Everglades ecosystem

- Avoid further destruction or degradation of these natural systems.
- Reestablish the ecological functions of these natural systems in areas where these functions have been damaged.
- Improve the overall management of water, fish and wildlife and recreation.
- Successfully restore and preserve these unique areas.

*Source*: **A.** "Save Our Everglades," Issue Paper (Tallahassee: Governor's Office, August 9, 1983). **B.** Executive Order 83–178 (Tallahassee: Governor's Office, November 1983).

---

## DOCUMENT 132: Edward O. Wilson on the Need for Conservation and a Conservation Ethic (1984, 1998)

Edward O. Wilson, one of the world's leading spokesmen for biodiversity and a professor and curator of entomology at the Museum of Comparative Zoology at Harvard University, decries "the folly our descendants are least likely to forgive us"—the devastation of the earth's resources. A neo-Malthusian, Wilson believes that the maintenance of the earth's biological diversity is essential for human physical and emotional well-being. He has predicted that the twenty-first century will be "the century of the environment," when humans will be forced to look at themselves "closely as a biological as well as a cultural species."[6]

## A. From *Biophilia*, 1984

The future of the conservation movement depends on . . . an advance in moral reasoning. Its maturation is linked to that of biology and a new hybrid field, bioethics, that deals with the many technological advances recently made possible by biology. Philosophers and scientists are applying a more formal analysis to such complex problems as the allocations of scarce organ transplants, heroic but extremely expensive efforts to prolong life, and the possible use of genetic engineering to alter human heredity. They have only begun to consider the relationships between human beings and organisms with the same rigor. It is clear that the key to precision lies in the understanding of motivation . . . why, say, they prefer a city with a park to a city alone. The goal is to join emotion with the rational analysis of emotion in order to create a deeper and more enduring conservation ethic.

. . .

[A] healthful environment, the warmth of kinship, right-sounding moral strictures, sure-bet economic gain, and a stirring of nostalgia and sentiment are the chief components of the surface ethic. Together they are enough to make a compelling case to most people most of the time for the preservation of organic diversity. But this is not nearly enough: every pause, every species allowed to go extinct, is a slide down the ratchet, an irreversible loss for all. It is time to invent moral reasoning of a new and more powerful kind, to look to the very roots of motivation and understand why, in what circumstances and on which occasions, we cherish and protect life. The elements from which a deep conservation ethic might be constructed include the impulses and biased forms of learning loosely classified as biophilia. Ranging from awe of the serpent to the idealization of the savanna and the hunter's mystique.

## B. From *Consilience*, 1998

[T]he global population is precariously large, will grow another third by 2020, and climb still more before peaking sometime after 2050. Humanity is improving per capita production, health, and longevity. But it is doing so by eating up the planet's capital, including irreplaceable natural resources. Humankind is approaching the limit of its food and water supply. As many as a billion people, moreover, remain in absolute poverty, with inadequate food from one day to the next and little or no medical care. Unlike any species that lived before, *Homo sapiens* is also changing the world's atmosphere and climate, lowering and polluting water tables, shrinking forests, and spreading deserts. It is extinguishing a large fraction of plant and animal species, an irreplaceable loss that will be viewed as catastrophic by future generations. Most of the stress originates directly or indirectly from a handful of industrialized countries. Their proven formulas are being eagerly adopted by the rest of the world. The emulation cannot be sustained, not with the same levels of consumption and waste. Even if the industrialization of developing countries is only partly successful, the environmental after shock will dwarf the population explosion that preceded it.

Source: **A.** Edward O. Wilson, *Biophilia* (Cambridge, MA: Harvard University Press, 1984), pp. 119, 138–41. **B.** Edward O. Wilson, *Consilience: The Unity of Knowledge* (New York: Knopf, 1998), p. 280.

---

# DOCUMENT 133: Jurgen Schmandt, Hilliard Roderick, and Andrew Morriss on Acid Rain and Friendly Neighbors (1985)

The earliest efforts to control industrial waste gases in order to reduce their negative impacts involved building taller smokestacks on factories. These tall stacks did nothing to reduce the amount of waste gases emitted; they simply spread the pollution farther afield. In the 1950s scientists began to suspect that certain waste gases produced by industrial activity—including sulfur and nitrogen oxides—not only harmed the human respiratory system but also caused damage to crops and the natural environment when they precipitated out of the air and settled on people, trees, and other living things. Federal efforts to control the production of the noxious gases that caused acid precipitation were initiated with the Clean Air Act of 1955 [see Document 93], but at the time the chemistry and mechanics of acid rain were not well understood. The National Acid Precipitation Assessment Program (NAPAP) was created in 1980 to provide data about the processes leading to acid precipitation and to evaluate its impact.

During these years Canada began to complain that acid rain resulting from industrial activity in the central United States was destroying forests in Canada, and it eventually became clear that bilateral cooperation would be necessary if measures to reduce acid rain were to prove effective.

Jurgen Schmandt, Hilliard Roderick, and Andrew Morriss offer a model for dealing with complex multinational environmental problems like acid rain, emphasizing that appropriate action must be taken on the national level if we want international environmental agreements to produce the desired changes.

All current initiatives [concerning acid rain]—in the United States, Canada, and Europe—have in common that they focus on one or two major pollutants ($SO_2$ and $NO_x$), and attempt to control acid rain under existing air pollution statutes. The existing policies were designed to control local air pollution. The proposed controls thus do not consider the fact that much of the danger of acid rain (for example, the damage to soils or drinking water) may result from the interaction of $SO_2$ and $NO_x$ with toxic pollutants and from complex chemical processes that occur during the long-range transport of the pollutants. All governments, in our view, need to broaden their view of acid rain and recognize the issue for what it is: a problem of unprecedented complexity, with many aspects that are not yet understood, with little precedent to guide action, and with powerful economic interests that see their livelihoods threatened. If that much is agreed upon, it becomes clear that what is needed is more than an expanded version of the current Clean Air Acts in Canada and the United States.

In the process of developing policy, it will help to increase the dialogue between policymakers and representatives of different interests. Canada and the United States share their environment. National policy in each country affects the other. Informal dialogue offers the opportunity for each nation to make its ideas and concerns known without the constraints of formal negotiations. But whatever will be achieved between the nations will have to be based on policy choices made at home. Without a sound acid rain policy at the domestic level, little can be accomplished internationally.

Past experience in addressing environmental disputes between Canada and the United States suggests that bilateral actions will play a useful but limited role; they are likely to supplement domestic initiatives but are unlikely to become the driving force for resolving the acid rain issue. We make the assumption, therefore, that no full-fledged international control policy will emerge, and that domestic-policy initiatives will have to lead the way. But within the framework of enlarged national policies, cooperation between the two countries (and eventually Mexico) must be agreed upon and implemented that far exceeds current political will, experience, and institutional capabilities. Specific measures include joint research, monitoring, control experiments affecting large areas, and harmonization of national policies.

Given the differences between Canada and the United States in size of population and gross national product, it is likely that decisions by the United States will determine the outcome of the acid rain issue. We expect that decisive domestic action will be delayed until the perceived damage is serious enough to generate broad support for another major initiative in environmental policy. Although such support seems to exist in Canada, the same is not yet the case in the United States. The fear of serious damage observed elsewhere has been the prime motivation for protection of environment and public health in the past, and this pattern is likely to continue in the case of acid rain. Once people become genuinely concerned about the effects of acid rain on wildlife, vegetation and human health, political momentum will build up fast.

*Source*: Jurgen Schmandt, Hilliard Roderick, and Andrew Morriss, "Acid Rain Is Different," Jurgen Schmandt and Hilliard Roderick, eds., *Acid Rain and Friendly Neighbors: The Policy Dispute between Canada and the United States* (Durham, NC: Duke University Press, 1985), pp. 19–20.

---

## DOCUMENT 134: United Nations Convention (1985) and Protocol (1987) on Ozone Depletion

Freons—chlorofluorocarbon gases (CFCs)—were developed in the 1930s to replace dangerous chemicals being used as commercial refrigerants. The invention of Freons led to widespread household use of refrigeration and the introduction of aerosol sprays. However, in 1974 the chemists Mario Molina and F. Sherwood Rowland reported that ultraviolet radiation from the sun causes the halogens chlorine and bromine to be released from halocarbons (compounds of carbon with halogens) and that these gases then combine with ozone in the stratosphere, resulting in a degradation of the ozone layer. The ozone layer protects life on earth from ultraviolet radiation from the sun.

Initially people scoffed at the suggestion that the halocarbons used in refrigerators, car air conditioners, and aerosol cans were a threat to the earth's ozone shield. Nevertheless, by 1978 chlorofluorocarbons had been banned from use in spray cans in the United States, and by 1983 there was a multinational call for a worldwide ban. Obviously, action by individual countries was inadequate to deal with the planet-wide ozone problem, and by the early 1980s it was widely recognized that only international action and global cooperation could stop ozone depletion.

The 1985 Vienna Convention for the Protection of the Ozone Layer provided a framework for dealing with ozone depletion on an international basis, and the 1987 Montreal Protocol spelled out precisely how the goals of the convention were to be achieved. The Montreal Protocol marked "the beginning of a new era of environmental statesmanship"[7] in which scientists, governmental leaders, and industrialists from around the world increasingly would have to work together to address complex issues posed by environmental threats and to formulate restrictions on industrial activity and economic development.

## A. Vienna Convention for the Protection of the Ozone Layer

Article 2: General Obligations

1. The Parties shall take appropriate measures in accordance with the provisions of this Convention and of those protocols in force to which they are party to protect human health and the environment against adverse effects resulting or likely to result from human activities which modify or are likely to modify the ozone layer.

2. To this end the Parties shall, in accordance with the means at their disposal and their capabilities:

(a) Co-operate by means of systematic observations, research and information exchange in order to better understand and assess the effects of human activities on the ozone layer and the effects on human health and the environment from modification of the ozone layer;

(b) Adopt appropriate legislative or administrative measures and co-operate in harmonizing appropriate policies to control, limit, reduce or prevent human activities under their jurisdiction or control should it be found that these activities have or are likely to have adverse effects resulting from modification or likely modification of the ozone layer.

Annex A

4. The following chemical substances of natural and anthropogenic origin, not listed in order of priority, are thought to have the potential to modify the chemical and physical properties of the ozone layer.

(a) Carbon substances

    (i)   Carbon monoxide (CO)

    (ii)  Carbon dioxide ($CO_2$)

    (iii) Methane ($CH_4$)

    (iv) Non-methane hydrocarbon species

(b) Nitrogen substances

    (i)   Nitrous oxide ($N_2O$)

    (ii)  Nitrogen oxides (NO)

(c) Chlorine substances

    (i)   Fully halogenated alkanes, e.g. $CCl_4$, $CFCl_3$ (CFC-11), $CF_2Cl_2$ (CFC-12), $C_2F_3Cl_3$ (CFC-113), $C_2F_4Cl_2$ (CFC-114)

    (ii)  Partially halogenated alkanes, e.g. $CH_3Cl$, $CHF_2Cl$ (CFC-22), $CH_3CCl_3$, $CHFCl_2$ (CFC-21)

(d) Bromine substances

    Fully halogenated alkanes, e.g. $CF_3Br$

(e) Hydrogen substances

    (i)   Hydrogen ($H_2$)

    (ii)  Water $H_2O$

B. Montreal Protocol on Substances That Deplete the Ozone Layer

Article 2: Control Measures

1. Each Party [to the Vienna Convention] shall ensure that for the twelve-month period commencing on the first day of the seventh month following the date of the entry into force of this Protocol, and in each twelve-month period thereafter, its calculated level of consumption of the controlled substances in Group I of Annex A does not exceed its calculated level of consumption in 1986. By the end of the same period, each Party producing one or more of these substances shall ensure that its calculated level of production of the substances does not exceed its calculated level of production in 1986, except that such level may have increased by no more than ten per cent based on the 1986 level. Such increase shall be permitted only so as to satisfy the basic domestic needs of the Parties operating under Article 5 [Special Situation of Developing Countries] and for the purposes of industrial rationalization [the transfer of all or a portion of the calculated level of production of one Party to another] between Parties.

*Source*: **A.** U.S. Department of State, "Protection of the Ozone Layer," Treaties and Other International Acts Series, no. 11097 (Washington, D.C.: Government Printing Office, March 22, 1985), pp. 5–6, 22–24. **B.** Montreal Protocol on Substances That Deplete the Ozone Layer (Nairobi: UNEP, September 1987), in Richard Benedick, *Ozone Diplomacy: New Directions in Safeguarding the Planet* (Cambridge, MA: Harvard University Press, 1991), Appendix B.

---

## DOCUMENT 135: United Church of Christ Commission for Racial Justice on Toxic Waste and Race (1987)

The United Church of Christ's report on the siting of hazardous waste facilities was the first comprehensive analysis of the relationship between hazardous waste sites and the racial, socioeconomic, and ethnic makeup of the communities in which they were located, and it aroused interest in the issue of environmental justice. The report was based on two studies: an analysis of the location of commercial hazardous waste facilities and a descriptive study of the racial composition of communities with uncontrolled toxic waste sites.

Until the late 1970's, most hazardous wastes were discarded without consideration of the dangers they posed. Moreover, proper care was lacking when hazardous chemicals were produced, stored and transported.

The glaring lack of hazardous waste management regulations created a permissive atmosphere for discarding wastes in the cheapest possible ways. The EPA recognized that, up to this time, 80 to 90 percent of hazardous wastes were disposed of without adequate safeguards for human health and the environment.

* * *

The descriptive study . . . found that more than half of the population in the United States lived in residential ZIP code areas with one or more uncontrolled toxic waste sites. The study also found that three out of every five Black and Hispanic Americans lived in communities with uncontrolled toxic waste sites.

* * *

The results of the study suggest that the disproportionate numbers of racial and ethnic persons residing in communities with commercial hazardous waste facilities is not a random occurrence, but rather a consistent pattern. Statistical association between race and the location of these facilities were stronger than any other association tested. The probability that this association occurred purely by chance is less than 1 in 10,000.

It is significant that race was consistently a more prominent factor in the location of commercial hazardous waste facilities than any other factor examined. This was clearly the case with respect to socio-economic status. The most striking relationship between socio-economic status and the location of commercial hazardous waste facilities was revealed after the study controlled for regional differences and urbanization. Household incomes and home values were substantially lower when communities with hazardous waste facilities were compared to communities in the surrounding area without such facilities. Mean household income was $2,745 less and the mean value of owner-occupied homes was $17,301 less.

* * *

It is clear from these studies that as the number of a community's racial and ethnic residents increases, the probability that some form of hazardous waste activity will occur also increases. The implications of that conclusion are serious. The Heckler Report[8] has detailed the excess deaths of Blacks and other racial and ethnic persons in this country; the presence of hazardous waste sites only serves to compound this problem. Since many facilities and uncontrolled sites tend to be located in those urban areas where large numbers of racial and ethnic Americans reside, the

potential risk caused by transportation spills, explosions, toxic emission, and groundwater contamination strikes hardest at racial and ethnic Americans who have been documented to be the most "at risk" when it comes to health and well-being.

*Source*: Commission for Racial Justice, United Church of Christ, *Toxic Wastes and Race in the United States: A National Report on the Racial and Socio-Economic Characteristics of Communities with Hazardous Waste Sites* (New York: Public Data Access, 1987), pp. 3, 13, 15–16, 17.

## DOCUMENT 136: Dixy Lee Ray Asks, "Who Speaks for Science?" (1990)

Although Dixy Lee Ray, the former governor of Washington and chair of the Atomic Energy Commission, recognizes that we must curtail the rapid depletion of natural resources and prevent pollution, she questions the placing of stringent environmental limitations on industrial development. Like Bernard Cohen [see Document 130], Ray is concerned about how the environmental riskiness of various activities, including industrial activities, is determined and how the perception of risk affects U.S. industrial and energy policy.

It is now widely accepted by the press and consequently by much of the general public that man's industrial activities are "fouling our nest" and pose a threat to the life of planet Earth, a threat that grows more ominous year by year. Is this conventional wisdom correct?

The risk one runs in challenging so widely held a belief is the risk of being judged an apologist for industry, or worse, to be accused of favoring pollution. Now my disclaimer: I am not in the pay of nor am I employed by any industry and I am as much opposed to pollution as anyone. But I do part company with alarmists who misuse science to foment fear and who clamor with increasing stridency that industrial progress must stop or be redirected into uneconomic alternatives because the world is going to pot. Is it?

\* \* \*

What are our real environmental concerns? Cancer-causing chemicals? Radiation, including radon? Carbon dioxide, ozone, the "greenhouse effect"? . . .

First, the cancer-causing chemicals. With the exception of childhood leukemia—always tragic, but relatively rare—cancer is a malady that

afflicts predominantly older adults and the aged. For most cancers, and there are many different kinds, the causes are complex, interactive, and often include genetic factors. If we look at the fatality records, the facts show that the total of carcinogenic substances targeted by the EPA—including chemicals in the work place, environment, food additives, and industrial products—cause *fewer than eight percent of all cancer deaths in America.*

The best scientific evidence points to genetics, viruses, sexual practices, diet, alcohol, and more than anything else, tobacco, as accounting for nearly all of the remaining 92 percent. Yet, the public, through constantly reported innuendo against industrial chemicals and radiation, is encouraged to believe otherwise. Moreover, a proper look at cancer statistics shows that, aside from a sharp increase in lung cancer caused by cigarette smoking, there have been no significant increases in the rate at which people die from any of the common forms of cancer over the last 50 years. In fact, there have been significant decreases in some types of cancer—for example, stomach cancer—during these decades of rapid industrialization and the introduction of many new man-made chemicals.

But most people believe cancer is caused by toxic substances created by industry. Why? Because they listen to the wrong spokesmen, and that is all they hear. National television has elevated sob-sister journalism to a new dramatic high, with emotional, heart-rending stories about cases of childhood leukemia and other individual or family tragedies as if they were epidemic. These stories capture public attention and play on natural sympathy, and these reactions, in turn, affect the decisions and budgets of governmental scientific agencies. In an internal memo, the Environmental Protection Agency (EPA) admits, with remarkable candor, "Our priorities [in regulating carcinogens] appear [to be] more closely aligned with public opinion than with our estimated risks." And with scientific evidence, too, I hasten to add!

. . . [R]adon has become a national health problem because of our well meant but stupid insistence on sealing up our homes and businesses to conserve energy, without considering the possible ill effects.

As to the "greenhouse effect," it's true that the concentration of carbon dioxide in the atmosphere has been increasing. It is also true that the *rate* of the carbon dioxide increase—and methane, hydrocarbons, sulfur and nitrogen oxides, and a few other substances—is now approximately one percent a year. Since increases of carbon dioxide have also occurred in the geological past, without the help of human industry, it is unclear whether the burning of fossil fuel is the preeminent or only cause of the present increase, however much it may be adding to the current totals. Moreover, it is not known what the consequences may be, if any, of this increase, nor how long it may last. But this does not stop the doomsayers from hypothesizing radical transformations and other adverse effects in the future.

We do *not* know what caused severe climatic changes in the geological past, but we can be sure they were *not* due to human industrial activity. Most likely, the causes were and still are colossal cosmic forces, quite outside human ability to control. Now that we live in an industrial, high technology society, there is no reason to believe that such cosmic forces have ceased to exist.

\* \* \*

The course of public events, especially in nuclear science and now increasingly in the chemical industry as well, has demonstrated over the last 10 to 15 years that scientists and engineers who speak on behalf of nuclear power and the chemical industry are not trusted. The public does not distinguish the Natural Resources Defense Council from the National Academy of Sciences and is far more likely to believe the opponents of science and technology than the supporters.

*Source*: Dixy Lee Ray with Lou Guzzo, *Trashing the Planet: How Science Can Help Us Deal with Acid Rain, Depletion of the Ozone, and Nuclear Waste (Among Other Things)* (New York: Regnery Publishing, Inc., 1990), pp. 3, 5–7.

## DOCUMENT 137: Roger Smith on Industry and the Environment (1990)

Research engineers at General Motors and other car and electronics companies have been working to develop a viable electric car for years. Public reception of a prototype GM electric car displayed at the January 1990 Automobile Show in Los Angeles encouraged GM president Roger Smith to commit GM to producing a commercial electric car. The impact of his commitment was more substantial than he would have believed, affecting not only GM but also prompting the California Air Resources Board to issue stringent vehicle emissions standards [see Document 138]. Unfortunately, the commercial version of the GM electric car produced in December 1996 proved to be too expensive and to have a battery capacity that was too limited to attract many buyers or leasers.

[E]nvironmental consciousness has moved from a fringe phenomenon ... into the mainstream value system of people everywhere in the developed world.... [T]his, of course, is of keen interest to us in the auto industry because it focuses attention directly on us and our products. And it bears directly on our two tasks—being successful in the market-place *and* fulfilling society's expectations. And really, the two are inter-related—because we can't be successful in the marketplace unless we fulfill society's expectations, and we won't have the resources to benefit society unless we're successful in the marketplace. What it all comes down to is, we have to market products that consumers want *and* that are environmentally sound. . . .

The first half of the title of my remarks is "Industry and the Environ-ment." What's the relationship between the two? Let me be very explicit about this: our job is to provide goods and services, which, along with their intrinsic purpose, do everything possible to promote cleaner air, water, and earth. You know, its a little ironic that we've been accused of not understanding the environmental implications of what we do. The truth is that no one understands the importance of the environment *better* than we. Our factories and our products are intimately related to the planet and all its vast resources; they depend on it in order to keep running. We've got to keep that in mind, if we're to preserve the jobs and livelihoods of our people, the health of our businesses—and ulti-mately, the benefits that customers derive from our products. And we *do* keep it in mind. Lots of companies have a long history of environ-mental consciousness and involvement—and General Motors is proud to be among them. As far back as 1955, we were conducting research on the atmosphere. Back in 1959, before there was an EPA, we discovered that crankcase emissions contribute to air pollution, and we proposed control measures that are in use on most vehicles today. I'm pleased to tell you—unequivocally—that over the years, GM has done more for automotive emissions control—we've come up with more patents, more publications, more research, more technical contributions—than any other auto company, anywhere.

In emissions and fuel-efficiency, our factories and our vehicles have made enormous strides—here and around the world. Just compare to-day's cars and trucks with those of 1970, the year of the first Earth Day. Today's vehicles have been downsized and built with lighter-weight ma-terials—so their average weight is down by 27%. We've converted most of them to front-wheel-drive, improved their aerodynamic efficiency by one-third, and set new specifications that cut the rolling resistance of our tires by half. We've reduced friction between parts, and put in more efficient transmissions and air-conditioning, as well as electronically-controlled, fuel-efficient engines. All of this—and more—has increased the fuel economy of GM cars, since the mid-70's, by 130%. And since clean air became a national goal, we've made big reductions in the ex-haust emissions from our cars and trucks: nitrogen oxides are down by 76%, and hydrocarbons and carbon monoxide are down by 96%. That's due in large part to the catalytic converter—a GM invention that *The Washington Post* recently called "the nation's most powerful weapon against urban smog." We started putting them into our 1975 cars, and just a few days ago, we marked the manufacture of 100 million of them.

\* \* \*

Another promising approach is the electric-powered car, which, I'm convinced, will play an important role in meeting our country's transportation needs and environmental goals. At the beginning of this year, General Motors introduced the Impact, a prototype electric car, to the public. It's the latest step in our continuing program to develop electric cars for personal transportation. . . . [O]ur goal is to be the first automobile company to mass-produce a modern, electric car. And not just *any* electric car, but one that's as safe as any vehicle on the road today, that has a range and cost that consumers will find appealing, and that performs as well as current internal combustion engine vehicles.

*Source*: Roger B. Smith, "Industry and the Environment: New Directions for the 90s," address to the National Press Club, Washington, D.C., April 18, 1990. GM Business Research Library, Detroit, MI.

## DOCUMENT 138: California Air Resources Board Lowers Its Vehicle Emissions Standards (1990, 1996)

The California Air Resources Board (CARB) was created in 1968 by Governor Ronald Reagan as an independent state agency to establish pollution emissions standards for air polluters in California, including automobiles and oil refineries. In the 1980s, as president of the United States, Reagan headed what many view as the most antienvironmental administration of the latter half of the twentieth century.

For years California's air pollution problem had been substantial, in part because the state's climate and geography work together to help produce what the Environmental Protection Agency (EPA) calls "nonattainment" conditions (meaning that the EPA standards for air quality are not met). Cities in California have an average of 109 nonattainment days per year, while cities and areas of high pollution situated on the other side of the Rocky Mountains generally have fewer than 50 nonattainment days per year.

The CARB vehicle emissions standards set in September 1990, whose strength may have been encouraged by GM president Roger Smith's pronouncements earlier in the year [see Document 137], proved difficult to meet, however, and in 1996 they were replaced by weaker standards. This is but one of many instances around the nation in which environmental goals set by federal, state, and local agencies and legislatures during the past three decades have been downgraded or ignored. Retreat from strict environmental standards has often occurred when the public or industry has deemed the cost of needed technology to be too high or the change in business activity or lifestyle too great.

### A. Original CARB Proposal for Vehicle Emissions Standards, September 1990

More than three-fourths of California's residents live in areas that do not meet at least one state or federal ambient air quality standard. Many areas of the state, including the South Coast, Bay Area, Central Valley, and Central Coast are non-attainment areas for the state ozone standard. Mobile sources are the largest contributors of precursors that react in the atmosphere to form ozone. These ozone precursors include hydrocarbons (or organic gases) and oxides of nitrogen ($NO_x$). Atmospheric ozone and other pollutants emitted directly by vehicles, including carbon monoxide (CO) and particulate matter (PM), are associated with respiratory irritation and illness. Motor vehicles also emit a substantial portion of both known and potential toxic contaminants. Vehicle-related toxics include benzene, a known human carcinogen, as well as potential carcinogens such as 1,3 butadiene, formaldehyde, acetaldehyde and diesel particulate.

. . .

Implementation of the proposed regulations would benefit air quality by reducing vehicle emissions throughout the state. The staff estimates that by 2010, the regulations would:

- reduce vehicular emissions of non-methane organic gases (NMOG) by 28 percent and $NO_x$ by 18 percent;
- reduce vehicular emissions of CO by 8 percent and PM by about 2 percent;
- reduce vehicular toxics (benzene, 1,3-butadiene, formaldehyde, acetaldehyde, and diesel particulate) associated with reactive organic gases and the associated potential cancer burden by about 20 to 40 potential cases per year statewide by 2010; and
- possibly reduce emissions of compounds identified as greenhouse gases that may contribute to global warming.

Beginning in the mid-1990's, vehicle manufacturers would be required to produce vehicles that meet TLEV [transitional limited emission vehicle], LEV [limited emission vehicle], ULEV [ultra-low emission vehicle],

**Table 1**
**Implementation Rates for Conventional Vehicles, TLEVs, LEVs, ULEVs, and ZEVs Used to Calculate Fleet Average Standards for Passenger Cars**

| Model Year | 0.39 | 0.25 | TLEV 0.125 | LEV 0.075 | ULEV 0.040 | ZEV* 0.00 | Fleet Avg. Standard [grams per mile emitted] |
|---|---|---|---|---|---|---|---|
| 1994 | 10% | 80% | 10% | | | | 0.250 |
| 1995 | | 85% | 15% | | | | 0.231 |
| 1996 | | 80% | 20% | | | | 0.225 |
| 1997 | | 73% | | 25% | 2% | | 0.202 |
| 1998 | | 48% | | 48% | 2% | 2% | 0.157 |
| 1999 | | 23% | | 73% | 2% | 2% | 0.113 |
| 2000 | | | | 96% | 2% | 2% | 0.073 |
| 2001 | | | | 90% | 5% | 5% | 0.070 |
| 2002 | | | | 85% | 10% | 5% | 0.068 |
| 2003 | | | | 75% | 15% | 10% | 0.062 |

*The percentage requirements for ZEVs are mandatory.

or ZEV [zero emission vehicle] exhaust emission standards. The emission standards for these vehicle categories differ in stringency, ranging from a 50 percent reduction in hydrocarbon emissions for TLEVs (compared to conventional gasoline-fueled vehicles sold in the same period) to allowing no emissions of any pollutants for ZEVs.

[The schedule for implementation of vehicle quotas (Table 1) shows the percentage of the total vehicles sold each year that will be required to meet the new standards and the level of pollution reduction they will need to reach it, as measured by the grams per mile of nonmethane organic gases released by the vehicles.]

### B. Proposed Changes to the CARB Vehicle Emissions Standards, January 1996

Based on information gathered through [a series of] public forums and [a] Battery [Technical Advisory] Panel, the staff [of CARB] is proposing to amend the LEV regulations to eliminate the percentage ZEV requirements for model years 1998 through 2002. The ten percent requirement for the 2003 model year would remain unchanged. This modification would allow auto manufacturers more time to develop and demonstrate ZEVs powered by advanced batteries and flexibility to determine the best time to introduce this new technology to the market. To encourage the

early production of advanced ZEVs, the staff is also proposing to add a provision to allow multiple credits for longer-range ZEVs produced prior to the 2003 model year. These ZEV credits could be applied to a manufacturer's 2003 and subsequent model year requirements.
    . . .

Under Title II of the Federal Clean Air Act (FCAA), the U.S. Environmental Protection Agency has promulgated comprehensive regulations to control emissions from new motor vehicles. . . . While both the federal and California automotive emissions standards are similar in purpose and scope, California has adopted standards that are generally more stringent and effective in order to address the severity of California's air pollution problem.

*Source*: **A**. State of California Air Resources Board, "Proposed Regulations for Low-Emission Vehicles and Clean Fuels," State of California Air Resources Boards staff report, August 13, 1990, pp. 3, 4, 24. **B**. California Air Resources Board, "Notice of Public Hearing to Consider Amendments to the Zero-Emission Vehicle Requirements for Passenger Cars and Light-Duty Trucks," January 30, 1996, p. 2.

---

## DOCUMENT 139: John P. Holdren on Energy and Human Well-Being (1990)

---

John P. Holdren is concerned about the failure of the United States to establish an energy policy that would encourage the development of clean, safe fuels to satisfy the nation's long-term energy needs, and he foresees negative social, political, economic, and environmental consequences from the continued failure to do so. A professor of energy and resources at the University of California, Berkeley, Holdren served as a scientific adviser to the Clinton administration as it prepared to negotiate the Kyoto Protocol [Document 142B].

[C]ivilization is not running out of energy resources in an absolute sense, nor is it running out of technological options for transforming these resources into the particular forms that our patterns of energy use require. We are, however, running out of the cheap oil and natural gas that powered much of the growth of modern industrialized societies, out of environmental capacity to absorb the impacts of burning coal, and out of public tolerance for the risks of nuclear fission. We seem to be lacking as well the commitment to make coal cleaner and fission safer, the money and endurance needed to develop long-term alternatives, the astuteness to embrace energy efficiency on the scale demanded and the consensus needed to fashion any coherent strategy at all.

These deficiencies suggest that civilization has entered a fundamental transition in the nature of the energy-society interaction without any collective recognition of the transition's character or its implications for human well-being. The transition is from convenient but ultimately scarce energy resources to less convenient but more abundant ones, from a direct and positive connection between energy and economic well-being to a complicated and multidimensional one, and from localized pockets of pollution and hazard to impacts that are regional and even global in scope.

The subject is also being transformed from one of limited political interest within nations to a focus of major political contention between them, from an issue dominated by decisions and concerns of the Western world to one in which the problems and prospects of all regions are inextricably linked, and from one of concern to only a small group of technologists and managers to one where the values and actions of every citizen matters.

Understanding this transition requires a look at the two-sided connection between energy and human well-being. Energy contributes positively to well-being by providing such consumer services as heating, lighting and cooking as well as serving as a necessary input to economic production. But the costs of energy—including not only the money and other resources devoted to obtaining and exploiting it but also the environmental and sociopolitical impacts—detract from well-being.

*Source*: From John P. Holdren, "Energy in Transition," *Scientific American* 263, no. 3 (September 1990): 157. Copyright © 1990 by Scientific American, Inc. All rights reserved.

---

## Document 140: Barry Lopez on a Sense of Place (1990)

Like John Steinbeck, Edward Abbey, and many Native Americans [see Documents 32, 80, 84, and 121], Barry Lopez, who writes about nature and human society, decries the greed that has been the root of the plunder of America's resources since Columbus's landing and whose end result has been the destruction of the places to which its inhabitants are spiritually bound.

A sense of place must include, at the very least, knowledge of what is inviolate about the relationship between a people and the place they occupy, and certainly, too, how the destruction of this relationship, or the failure to attend to it, wounds people. . . .

If, in a philosophy of place, we examine our love of the land—I do not mean a romantic love, but the love Edward Wilson calls biophilia, love of what is alive, and the physical context in which it lives, which we call "the hollow" or "the cane brake" or "the woody draw" or "the canyon"—if, in measuring our love, we feel anger, I think we have a further obligation. It is to develop a hard and focused anger at what continues to be done to the land not so that people can survive, but so that a relatively few people can amass wealth.

\* \* \*

One of our deepest frustrations as a culture, I think, must be that we have made so extreme an investment in mining the continent, created such an infrastructure of nearly endless jobs predicated on the removal and distribution of trees, water, minerals, fish, plants, and oil, that we cannot imagine stopping. In the part of the country where I live [the Pacific Northwest], thousands of men are now asking themselves what jobs they will have—for they can see the handwriting on the wall—when they are told they cannot cut down the last few trees and that what little replanting they've done—if it actually works—will not produce enough timber soon enough to ensure their jobs.

The frustration of these men, who are my neighbors, is a frustration I am not deeply sympathetic to—their employers have behaved like wastrels, and they have known for years that this was coming. But in another way I am sympathetic, for these men are trying to live out an American nightmare which our system of schools and our voices of government never told them was ill-founded. There is not the raw material in the woods, or beyond, to make all of us rich. And in striving for it, we will only make ourselves, all of us, poor.

*Source*: Barry Lopez, *The Rediscovery of North America* (New York: Vintage Books/ Random House, 1992), pp. 40–42, 44–46.

## DOCUMENT 141: Albert Gore on the Ecological Perspective (1992)

Albert Gore, a former senator from Tennessee and vice president in the Clinton administration, has long been concerned about the environment. This selection comes from a book on the environment written at about the time Bill Clinton and Al Gore began their first campaign for election to the White House.

The edifice of civilization has become astonishingly complex, but as it grows ever more elaborate, we feel increasingly distant from our roots in the earth. In one sense, civilization itself has been on a journey from its foundations in the world of nature to an ever more contrived, controlled, and manufactured world of our own imitative and sometimes arrogant design. And in my view, the price has been high. At some point during this journey we lost our feeling of connectedness to the rest of nature. We now dare to wonder: Are we so unique and powerful as to be essentially separate from the earth?

Many of us act—and think—as if the answer is yes. It is now all too easy to regard the earth as a collection of "resources" having an intrinsic value no larger than their usefulness at the moment. Thanks in part to the scientific revolution, we organize our knowledge of the natural world into smaller and smaller segments and assume that the connections between these separate compartments aren't really important. In our fascination with the parts of nature, we forget to see the whole.

The ecological perspective begins with a view of the whole, an understanding of how the various parts of nature interact in patterns that tend toward balance and persist over time. But this perspective cannot treat the earth as something separate from human civilization; we are part of the whole too, and looking at it ultimately means also looking at ourselves. And if we do not see that the human part of nature has an increasingly powerful influence over the whole of nature—that we are, in effect, a natural force just like the winds and the tides—then we will not be able to see how dangerously we are threatening to push the earth out of balance.

Our perspective is badly foreshortened in another way as well. Too often we are unwilling to look beyond ourselves to see the effect of our actions today on our children and grandchildren. I am convinced that many people have lost their faith in the future, because in virtually every facet of our civilization we are beginning to act as if our future is now so much in doubt that it makes more sense to focus exclusively on our current needs and short-term problems. This growing tendency to discount the value of investments made for the long term—whether of wealth, effort, or caution—may have begun with the realization that nuclear weaponry had introduced a new potential for an end to civilization. But whatever its genesis, our willingness to ignore the consequences of our actions has combined with our belief that we are separate from nature to produce a genuine crisis in the way we relate to the world around us.

*Source*: Al Gore, *Earth in the Balance* (Boston: Houghton Mifflin, 1992), pp. 1–2.

## DOCUMENT 142: United Nations Convention (1992) and Protocol (1997) on Climate Change

By the mid-1980s some scientists were beginning to suspect that heat-trapping gases such as carbon dioxide, emitted during the burning of fossil fuels, and methane were not only causing acid rain and thinning the ozone layer but were also effecting a change in the earth's climate. They proposed that just as the windows of a greenhouse allow heat from the sun to pass through them and then hold that heat inside, thereby raising the interior temperature, so too do heat-trapping gases permit some of the sun's radiant energy to penetrate the atmosphere all the way to the earth's surface and then prevent the resulting heat from escaping the earth's atmosphere, thereby causing a warming of the earth.

In order to evaluate the scientific basis of this theory, consider the potential impact of global warming, and examine possible policy responses to climate change, the Intergovernmental Panel on Climate Change (IPCC), an international panel of scientists, government representatives, and policymakers working in three groups, was established in 1988. The 1990 report of the IPCC's scientific group[9] concluded that the concentration of greenhouse gases in the earth's atmosphere was increasing and that this might be contributing to a change in the earth's climate.

While some environmentalists, such as Bill McKibben, worry that as a result of human activity "the Antarctic ice sheet could fail more quickly than previously believed" and that we are headed for disaster unless we act immediately to stop the warming trend,[10] there is no absolute proof that global warming is occurring or that greenhouse gases are the cause of the apparent warming trend. Indeed, some people believe that higher temperatures recorded around the world in recent years merely reflect normal long-term fluctuations in the global climate. However, many environmentalists, scientists, and policymakers feel that the nations of the world would be derelict if they did not take action to forestall a highly probable imminent catastrophe.

The United Nations Framework Convention on Climate Change, adopted in May 1992, was the international community's response to these concerns. Its goal was to address all the greenhouse gas issues not covered by the Montreal Protocol [Document 134B]. The details of how the Climate Change Convention was to be implemented were set forth in the Kyoto Protocol. This historic agreement imposed legally binding limits on the production by developed countries of man-made greenhouse gases. Although the United States was one of the first countries to sign the Climate Change Convention, it was one of the last to sign the Kyoto Protocol. The business community—represented here by the Business Roundtable, which includes the chief executives of more than two hundred of the largest corporations in the United States—objects to many of the details of this unprecedented agreement and opposes its ratification in its present form.

## A. United Nations Framework Convention on Climate Change

*The parties to this Convention,*

*Acknowledging* that change in the Earth's climate and its adverse effects are a common concern of humankind,

*Concerned* that human activities have been substantially increasing the atmospheric concentrations of greenhouse gases, that these increases enhance the natural greenhouse effect, and that this will result on average in an additional warming of the Earth's surface and atmosphere and may adversely affect natural ecosystems and humankind,

*Noting* that the largest share of historical and current global emissions of greenhouse gases has originated in developed countries, that per capita emissions in developing countries are still relatively low and that the share of global emissions originating in developing countries will grow to meet their social and development needs,

*Aware* of the role and importance in terrestrial and marine ecosystems of sinks and reservoirs of greenhouse gases,

*Noting* that there are many uncertainties in predictions of climate change, particularly with regard to the timing, magnitude and regional patterns thereof,

*Acknowledging* that the global nature of climate change calls for the widest possible cooperation by all countries and their participation in an effective and appropriate international response, in accordance with their common but differentiated responsibilities and respective capabilities and their social and economic conditions . . .

*Affirming* that responses to climate change should be coordinated with social and economic development in an integrated manner with a view to avoiding adverse impacts on the latter, taking into full account the legitimate priority needs of developing countries for the achievement of sustained economic growth and the eradication of poverty . . .

*Have agreed as follow:*

. . .

The ultimate objective of this Convention and any related legal instruments that the Conference of the Parties may adopt is to achieve . . . stabilization of greenhouse gas concentrations in the atmosphere at a level that would prevent dangerous anthropogenic interference with the climate system. Such a level should be achieved within a time-frame sufficient to allow ecosystems to adapt naturally to climate change, to ensure that food production is not threatened and to enable economic development to proceed in a sustainable manner.

## B. Kyoto Protocol

Article 3

1. The Parties included in Annex I [all the countries of Western Europe and most of the countries of Eastern Europe, as well as the United States, Canada, Japan, Australia, and New Zealand] shall, individually or jointly, ensure that their aggregate anthropogenic carbon dioxide equivalent emissions of the greenhouse gases listed in Annex A [carbon dioxide, methane, nitrous oxide, hydrofluorocarbons, perfluorocarbons, and sulphur hexafluoride] do not exceed their assigned amounts, calculated pursuant to their quantified emission limitation and reduction commitments . . . , with a view to reducing their overall emissions of such gases by at least 5 percent below 1990 levels in the commitment period 2008 to 2012.

2. Each Party included in Annex I shall, by 2005, have made demonstrable progress in achieving its commitments under this Protocol.

6. [A] certain degree of flexibility shall be allowed by the Conference of the Parties . . . to the Parties included in Annex I undergoing the process of transition to a market economy.

Article 6

1. For the purpose of meeting its commitments under Article 3, any Party included in Annex I may transfer to, or acquire from, any other such Party emission reduction units resulting from projects aimed at reducing anthropogenic emissions by sources or enhancing anthropogenic removals by sinks of greenhouse gases in any sector of the economy.

Article 12

3. Under the clean development mechanism:

(a) Parties not included in Annex I will benefit from project activities resulting in certified emission reductions; and

(b) Parties included in Annex I may use the certified emission reductions accruing from such project activities to contribute to compliance with part of their quantified emission limitation and reduction commitments under Article 3.

## C. The Business Roundtable's Key Issues of Concern about the Kyoto Protocol, 1998

- **The targets and timetables would require the U.S. to make significant and immediate cuts in energy use**. The Protocol would require the U.S. to reduce emissions 7% below 1990 levels by 2009–2012, an unprecedented 41% reduction in projected emission levels. The process of Senate ratification and the subsequent lengthy domestic implementation process post-ratification would leave the U.S. very little time to make the painful choices regarding energy use that will be necessary to achieve these reductions. In addition, because the Protocol sets different targets for each industrialized country and the target is based on what is now an eight-year-old baseline, the U.S. in effect will shoulder a disproportionate level of reduction and may be placed at a competitive disadvantage.

- **Unless the Developing Countries also commit to emission reductions, the Protocol is incomplete and will not work.** The Byrd-Hagel Resolution, unanimously adopted by the U.S. Senate in July 1997, states that the U.S. should not be a signatory to any protocol unless it mandates "new specific scheduled commitments to limit or reduce greenhouse gas emissions for the Developing Country Parties within the same compliance period." Many Developing Countries are rapidly growing their economies and will become the largest emitters of greenhouse gases in the next 15–20 years. *Greenhouse gases know no boundaries, and stabilization of greenhouse gas concentrations cannot be achieved without global participation in a limitation-reduction effort.* Moreover, regulating the emissions of only a handful of countries could lead to the migration of energy-intensive production—such as the chemicals, steel, petroleum refining, aluminum and mining industries—from the industrialized countries to the growing Developing Countries.

- **Certain carbon "sinks" may be used to offset emission reductions, but the Protocol does not establish how sinks will be calculated.** Carbon sinks, a natural system that absorbs carbon dioxide, have tremendous potential as a means of reducing emissions, but too much is currently unknown to make a fair determination. *It is unclear how sinks might help the U.S. reach its emission-reduction commitment.* It is understood that the rules for sinks will be addressed at the Fourth Conference of the Parties (COP4) in Buenos Aires, but until they are fully fleshed out their potential impact cannot be evaluated.

- **The Protocol contains no mechanisms for compliance and enforcement.** Simply put, it would be inappropriate for any country to ratify a legally binding international agreement that lacks compliance guidelines and enforcement mechanisms. The Protocol outlines a system of domestic monitoring with oversight by international review teams, but what constitutes compliance and who judges it will not be determined until after the Protocol enters into force. *The means of enforcement—also unknown—is equally critical, since a country's noncompliance could give it a competitive advantage over the U.S. and eviscerate the agreement's environmental goals.*

- **The Protocol includes flexible, market-based mechanisms to achieve emission reductions, but it does not establish how these mechanisms would work and to what extent they could be used.** The U.S. intends to rely heavily on market-based mechanisms to find the most efficient and cost-effective ways to reduce emissions. But until the rules and regulations are established it is uncertain how effective these mechanisms will be and to what extent they can be used by companies. Many countries are resisting these market-based mechanisms and their reluctance may hinder the development of adequate free-market guidelines. *The absence of many countries from the marketplace, and the possible limitations and restrictions on the marketplace, could render these mechanisms useless or of little value.*

- **The Protocol leaves the door open for the imposition of mandatory policies and measures to meet commitments.** Just as the U.S. favors flexible market mechanisms, the European Union and many Developing Countries favor harmonized, mandatory "command-and control" policies and measures—such as carbon taxes and CAFÉ standards—to meet commitments, and they will have numerous opportunities to seek adoption of these policies.

- **Finally, the procedures for ratification of, and amendment to, the Kyoto Protocol make it difficult to remedy before it enters into force.** The Protocol may not be amended, nor can rules and guidelines be adopted, until after the Protocol enters into force. The Clinton Administration is now considering the negotiation of a separate or supplemental protocol to attain necessary additional commitments, but this approach would open all issues to further negotiation.

**The Business Roundtable believes that the Congress and the American people cannot evaluate the Kyoto Protocol until the Administration sets out a plan as to how it intends to meet the targets of the Protocol.** To place the magnitude of the U.S. reduction commitments in perspective, it is the equivalent of having to eliminate all current emissions from either the U.S. transportation sector, or the utilities sector (residential and commercial sources), or industry. The Administration needs to detail how targets in the Protocol will be met, and how the burden will be distributed among the various sectors of the economy.

The Business Roundtable feels that a public dialogue must take place on the major issues highlighted in our Gap Analysis before the Protocol becomes the law of the land and government agencies begin to write regulations.

*Source*: **A**. *United Nations Framework Convention on Climate Change*, U.N. document A/AC.237/18, Part II/Add. 1 and Corr. 1, New York, May 9, 1992, pp. 1, 3, 4. **B**. *Kyoto Protocol to the United Nations Framework Convention on Climate Change*, U.N. document FCCC/CP/1997/L.7/Add. 1, Kyoto, December 1997, pp. 3, 4, 7, 12. **C**. The Business Roundtable Environmental Task Force, *The Kyoto Protocol: A Gap Analysis* (Washington, D.C.: The Business Roundtable, June 1998), pp. 1–3.

---

## DOCUMENT 143: Carl Safina on the Decline of Fishes (1995)

Carl Safina, director of the Audubon Society's Living Oceans program and author of *Song for the Blue Ocean* (1997), sees a connection between high technology and the decline in the world's fish populations.

In 1992 the United Nations banned large drift nets because they frequently trapped a wide range of marine life in addition to the sought-after fish; nevertheless, as Safina notes, they continue to be used by fisherman from several nations.

Agribusiness also poses a threat to the fishing industry. Nitrogen-rich runoff from land that has been heavily fertilized or contains large amounts of animal wastes has caused red tides—huge growths of small organisms on the surface of near-shore waters—that have resulted in dead zones because oxygen cannot penetrate them.

Because wild fish regenerate at rates determined by nature, attempts to increase their supply to the marketplace must eventually run into limits. That threshold seems to have been passed in all parts of the Atlantic, Mediterranean and Pacific. . . . Worldwide, the extraction of wild fish peaked at 82 million metric tons in 1989. Since then, the long-term growth trend has been replaced by stagnation or decline.

In some areas where the catches peaked as long ago as the early 1970s, current landings have decreased by more than 50 percent. Even more disturbingly, some of the world's greatest fishing grounds, including the Grand Banks and Georges Bank of eastern North America, are now essentially closed following their collapse—the formerly dominant fauna have been reduced to a tiny fraction of their previous abundance and are considered commercially extinct.

Recognizing that a basic shift has occurred, the members of the United Nations's Food and Agriculture Organization (a body that encouraged the expansion of large-scale industrial fishing only a decade ago) recently concluded that the operation of the world's fisheries cannot be sustained. They now acknowledge that substantial damage has already been done to the marine environment and to the many economies that depend on this natural resource.

* * *

How did this collapse happen? An explosion of fishing technologies occurred during the 1950s and 1960s. During that time, fishers adapted various military technologies to hunting on the high seas. Radar allowed boats to navigate in total fog, and sonar made it possible to detect schools of fish deep under the oceans' opaque blanket. Electronic navigation aids such as LORAN (Long-Range Navigation) and satellite positioning systems turned the trackless sea into a grid so that vessels could return to within 50 feet of a chosen location, such as sites where fish gathered and bred. . . .

Many industrial fishing vessels are floating factories deploying gear of enormous proportions: 80 miles of submerged longlines with thousands of baited hoods, bag-shaped trawl nets large enough to engulf 12 jumbo jetliners and 40-mile-long drift nets (still in use by some countries). Pressure from industrial fishing is so intense that 80 to 90 percent of the fish in some populations are removed every year.

. . . Fishers have countered loss of preferred fish by switching to species of lesser value, usually those positioned lower in the food web—a practice that robs larger fishes, marine mammals and seabirds of food. During the 1980s, five of the less desirable species made up nearly 30 percent of the world fish catch but accounted for only 6 percent of its monetary value.

* * *

[T]he development of aquaculture has not reduced the pressure on wild populations. Strangely it may do the opposite. Shrimp farming has created a demand for otherwise worthless catch because it can be used as feed. In some countries, shrimp farmers are now investing in trawl nets with fine mesh to catch everything they can for shrimp food, a practice known as biomass fishing. Much of the catch are juveniles of valuable species, and so these fish never have the opportunity to reproduce.

Fish farms can hurt wild populations because the construction of pens along the coast often requires cutting down mangroves—the submerged roots of these salt-tolerant trees provide a natural nursery for shrimp and fish.

*Source*: From Carl Safina, "The World's Imperiled Fish," *Scientific American* 273 (November 1995): 48–49. Copyright © 1995 by Scientific American, Inc. All rights reserved.

## DOCUMENT 144: Edward Tenner on Shifting Liability (1996)

Surely the scientists who developed DDT did not anticipate that the pesticide would drive the bald eagle and several other prized species to the brink of extinction. Neither did the builders of the electric power stations in the Ohio Valley anticipate that acid rain would result from the construction of the tall smokestacks that rid the region between Pittsburgh and Cincinnati of much of its pollution.

Edward Tenner, a historian and a visiting researcher in the Geosciences Department of Princeton University, questions whether many so-called scientific, technological, and medical advances of the twentieth century are not merely shifting the nature of environmental liability. While sixteenth-century scientists and philosophers such as Francis Bacon [see Document 8] believed that technology would give humans control over nature, in the late twentieth century it has become increasingly evident to more and more people that it is impossible to foresee all the consequences of a technological innovation[11] or to predict precisely how an ecosystem will respond to its manipulation by people.

Classic disasters were deterministic. Cause and effect were linked. An exploding boiler killed those it killed, and spared those it spared. Late-twentieth-century disasters are expressed as deviations from a baseline of "normal" background tragedy. The truth is not in immediate view. It emerges from the statistical inferences of trained professionals. . . . The old disasters were localized and sudden. New ones may be global and gradual, from radioactive isotopes in milk in the 1950s to climate change in the 1990s.

Our control of the acute has indirectly promoted chronic problems. Medical researchers have recognized this trend for years and have been shifting their efforts to chronic diseases—though so far not with the same results they have had with injury, infection, and acute illness. Our ability to transport animals and plants among continents, deliberately and accidentally, has on balance been decreasing rather than promoting species diversity. But the invaders have also failed to be as catastrophic to trees and crops as some had feared. Like many chronic illnesses, they have become manageable nuisances, neither conquerable nor fatal, but demanding time-consuming vigilance. Our efforts to modify our environment have also produced chronic problems: the comforts of home have helped produce the annoyances of allergies, suppressing forest fires has helped make them a greater threat, and protecting the shoreline is helping to erode it.

\* \* \*

By intensifying our protection against some forms of natural danger, we have sometimes only shifted greater liability to the future: a rearranging effect. We have traded acute problems for gradual but accumulating ones. This is especially true of the environmental disasters affecting energy.

*Source*: Edward Tenner, *Why Things Bite Back: Technology and the Revenge of Unintended Consequences* (New York: Knopf, 1996), pp. 24–25, 72.

## DOCUMENT 145: Jane Lubchenco on Environmental Issues for the Twenty-first Century (1997)

In an address to the American Association of the Advancement of Science (AAAS) in February 1997, Jane Lubchenco, a professor of zoology at Oregon State University who was then serving as president of the AAAS, discussed the impact of humans on the earth and the impact of the environment on human life.

[W]e now live on a human-dominated planet. The growth of the human population and the growth in amount of resources used are altering Earth in unprecedented ways. Through the activities of agriculture, fisheries, industry, recreation, and international commerce, humans cause three general classes of change. Human enterprises (i) transform the land and sea—through land clearing, forestry, grazing, urbanization, mining, trawling, dredging, and so on; (ii) alter the major biogeochemical cycles—of carbon, nitrogen, water, synthetic chemicals, and so on; and (iii) add or remove species and genetically distinct populations—via habitat alteration or loss, hunting, fishing, and introductions and invasions of species.

... [P. M.] Vitousek and colleagues[12] have provided a succinct and dramatic summary of the extent of human domination of Earth in the following six conclusions: (i) between one-third and one-half of the land surface has been transformed by human action; (ii) the carbon dioxide concentration in the atmosphere has increased by nearly 30% since the beginning of the Industrial Revolution; (iii) more atmospheric nitrogen is fixed by humanity than by all natural terrestrial sources combined; (iv) more than half of all accessible surface fresh water is put to use by humanity; (v) about one-quarter of the bird species on Earth have been driven to extinction; and (vi) approximately two-thirds of major marine fisheries are fully exploited, overexploited, or depleted.

\* \* \*

[D]uring the last few decades humans have emerged as a new force of nature. We are modifying physical, chemical, and biological systems in new ways, at faster rates, and over larger spatial scales than ever recorded on Earth. Humans have unwittingly embarked upon a grand experiment with our planet. The outcome of this experiment is unknown, but has profound implications for all of life on Earth.

* * *

As we begin to appreciate the intimate fashion in which humans depend on the ecological systems of the planet, it is becoming increasingly obvious that numerous issues that we have previously thought of as independent of the environment are intimately connected to it. Human health, the economy, social justice, and national security all have important environmental aspects whose magnitude is not generally appreciated.

*Source*: Excerpted with permission from Jane Lubchenco, "Entering the Century of the Environment: A New Social Contract for Science," *Science* 279 (January 23, 1998): 491–93. Copyright 1998 American Association for the Advancement of Science.

## Document 146: Jeremy Rifkin on Biotechnology and the Environment (1998)

Jeremy Rifkin, president of the Washington, D.C.-based Foundation on Economic Trends, is a futurist who writes about developments in science and technology and their impending influence on economic and social trends. His book *The Biotech Century* explores some of the possible effects on the environment of manipulating nature to produce overnight the kinds of changes that in the past required tens of thousands of years.

While the biotech revolution will reshape the global economy and remake our society, it is likely to have an equally significant impact on the Earth's environment. The new technologies of the Genetic Age allow scientists, corporations, and governments to manipulate the natural world at the most fundamental level—the genetic components that help orchestrate the developmental processes in all forms of life. In this regard, it is probably not overstating the case to suggest that the growing arsenal of biotechnologies is providing us with powerful new tools to engage in what will surely be the most radical experiment on the Earth's life forms and ecosystems in history. Imagine the wholesale transfer of genes between totally unrelated species and across all biological boundaries—plant, animal and human—creating thousands of novel life forms in a brief moment of evolutionary time. Then, with clonal propagation, mass-producing countless replicas of these new creations, releasing them into the biosphere to propagate, mutate, proliferate, and migrate, colonizing the land, water, and air.

* * *

Genetically engineered organisms differ from petrochemical products in several important ways. Because they are alive, genetically engineered organisms are inherently more unpredictable than petrochemicals in the way they interact with other living things in the environment. Consequently, it is much more difficult to assess all of the potential impacts that a genetically engineered organism might have on the Earth's ecosystems.

Genetically engineered products also reproduce. They grow and they migrate. Unlike many petrochemical products, it is difficult to constrain them within a given geographical locale. Once released, it is virtually impossible to recall genetically engineered organisms back to the laboratory, especially those organisms that are microscopic in nature. For all these reasons, genetically engineered organisms may pose far greater long-term potential risks to the environment than petrochemicals.

The risks in releasing novel genetically engineered organisms into the biosphere are similar to those we've encountered in introducing exotic organisms into the North American habitat. Over the past several hundred years, thousands of non-native organisms have been brought to America from other regions of the world. While many of these organisms have adapted to the North American ecosystems without severe dislocations, a small percentage of them have run wild, wreaking havoc on the flora and fauna of the continent. Gypsy moth, Kudzu vine, Dutch elm disease, chestnut blight, starlings, and Mediterranean fruit flies come easily to mind. . . . Each year the American continent is ravaged by these non-native organisms, with destruction to plant and animal life running into the billions of dollars.

Whenever a genetically engineered organism is released, there is always a small chance that it too will run amok because, like non-indigenous species, it has been artificially introduced into a complex environment that has developed a web of highly integrated relationships over long periods of evolutionary history. Each new synthetic introduction is tantamount to playing ecological roulette. That is, while there is only a small chance of it triggering an environmental explosion, if it does, the consequences could be significant and irreversible.

Global life-sciences companies are expected to introduce thousands of new genetically engineered organisms into the environment in the coming century, just as industrial companies introduced thousands of petrochemical products into the environment over the course of the past two centuries.

* * *

The reseeding of the planet with a laboratory-conceived second Genesis is likely to enjoy some enviable short-term market successes, only to ultimately fail at the hands of an unpredictable and noncompliant nature. While the genetic technologies we've invented to recolonize the biology of the planet are formidable, our utter lack of knowledge of the intricate workings of the biosphere we're experimenting on poses an even more formidable constraint. The introduction of new genetic-engineering tools and the opening up of global commerce allow an emerging "life industry" to "reinvent" nature and manage it on a worldwide scale. The new colonization, however, is without a compass. There is no predictive ecology to help guide this journey and likely never will be, as nature is far too alive, complex, and variable to ever be predictably modeled by scientists. We may, in the end, find ourselves lost and cast adrift in this artificial new world we're creating for ourselves in the Biotech Century.

*Source*: Jeremy Rifkin, *The Biotech Century: Harnessing the Gene and Remaking the World* (New York: Jeremy P. Tarcher/Putnam, 1998), pp. 67, 72–74, 115.

# Appendix I: Landmarks in U.S. Environmental Legislation

The following list includes precedent-setting state and federal legislation. Items marked with an asterisk appear as documents in the text.

| | |
|---|---|
| 1807 | Lead Mine Leasing Act establishes policy of leasing mineral rights; repealed in 1847. |
| 1817 | *Forest Preserve Act establishes first federal forest. |
| 1818 | *Bird Protection Law, Commonwealth of Massachusetts. |
| 1841 | Preemption Act allows settlers to squat on public lands and, if the land is put up for sale, to have the first right of purchase at $1.25 per acre. |
| 1850 | *Swamp and Overflow Act (Swamplands Act). |
| 1862 | *Homestead Act gives settlers free land on the condition that they inhabit it and cultivate it. |
| 1864 | *Act Granting Yo-semite to California creates first state park. |
| 1866 | General Mining Act establishes that public mineral lands should be free and open to exploration and occupation. |
| 1870 | Timber and Stone Act permits sale of uncultivatable public lands containing timber and stone but not minerals. |
| 1872 | *Mining Act allows private acquisition and exploration of public lands containing mineral deposits. |
| | *Act to Set Apart Land near the Head-Waters of the Yellowstone River as a Public Park creates first national park. |
| 1873 | Timber Culture Act gives land to individuals who agree to plant trees on a portion of it. |
| 1877 | Desert Land Act allows individuals to purchase federal lands cheaply on the condition that they begin irrigating the land within three years of date of purchase. |

| 1878 | Free Timber Act (Timber Cutting Act) sets rules for the acquisition of timber from federal public lands. |
| 1885 | *Act Establishing the Adirondack Forest Preserve, New York State. |
| 1891 | *Forest Reserve Act/Repeal of Timber Culture Acts sets aside 13 million acres in public domain as forest reserves (later to become national forests). |
| | Animal Inspection Act. |
| 1894 | Carey Act distributes federal lands to various states on condition that they irrigate the land. |
| 1897 | Forest Management Act defines purpose of forest reserves; amended in 1960 by Multiple Use Act. |
| 1899 | *Rivers and Harbors Act bans discharge of wastes into navigable rivers and harbors without permission from U.S. Army Corps of Engineers; first federal antipollution law. |
| 1900 | Lacey Act forbids interstate shipment of game killed in violation of state laws. |
| 1902 | *Reclamation Act (Newlands Act) launches a federal land reclamation program. |
| 1905 | Transfer Act puts 86 million acres under domain of Forest Service. |
| 1906 | Pure Food and Drug Act attempts to prevent adulteration and mislabeling of foods and drugs. |
| | Antiquities Act allows areas of scientific or historical interest on federal land to be set aside as national monuments. |
| 1907 | Meat Inspection Act requires inspection of meat-packing plants to eliminate unsanitary conditions; amended 1967. |
| 1910 | Insecticide Act prohibits interstate transport of mislabeled and adulterated insecticides. |
| 1911 | Weeks Act creates numerous national forests in the East. |
| 1913 | Migratory Bird Act (McClean Law) puts migratory birds under federal protection. |
| 1914 | Drinking Water Standards Act. |
| 1916 | National Parks Service Act establishes National Parks Service to manage national parks. |
| 1918 | Federal Migratory Bird Treaty Act implements 1916 U.S.-Canada treaty restricting the hunting of migratory birds. |
| 1920 | Federal Water Power Act. |
| | Mineral Leasing Act establishes regulations for mining on federal lands. |

| | |
|---|---|
| 1924 | Clarke-McNary Act extends power of federal government to purchase lands for national forest system and makes provision for private, state, and federal cooperation in forest management. |
| | Oil Pollution Act prohibits dumping of oil in navigable waters except in dire emergencies. |
| 1934 | Taylor Grazing Act regulates grazing on federal lands. |
| 1935 | Soil Conservation Act increases federal involvement in erosion control. |
| | Omnibus Flood Control Act. |
| 1937 | Pittman-Robertson Act promotes wildlife restoration. |
| 1938 | Food, Drug, and Cosmetics Act extends federal authority over misbranded foods, drugs, and cosmetics; amended 1958. |
| 1946 | Federal Lands Management Act consolidates administration of public lands. |
| 1947 | Federal Insecticide, Fungicide, and Rodenticide Act (FIFRA) attempts to protect consumers from fraudulent pesticide products; amended in 1972 to give greater focus to monitoring health and environmental consequences of pesticides. |
| 1948 | Federal Water Pollution Control Act (FWPA) authorizes Public Health Service to aid states in developing of water pollution control programs and planning sewage treatment plants; amended 1972. |
| 1953 | Outer Continental Shelf Lands Act (Submerged Lands Act) asserts that the federal government has authority over the development of mineral resources in the outer continental shelf; amended 1990. |
| 1955 | *Federal Clean Air Act (FCAA) (Air Pollution Control Act) provides aid for state and local air quality control and research programs; amended 1963, 1965, 1970,* 1977, 1990. |
| 1956 | Water Pollution Control Act creates water pollution control programs on interstate waterways. |
| 1957 | Poultry Products Inspection Act makes poultry inspection mandatory; amended 1968. |
| 1958 | Food Additive Amendment (Delaney Clause) to 1938 Food, Drug, and Cosmetics Act prohibits approval of any additive "found to induce cancer in man or animal." |
| 1960 | Federal Hazardous Substances Act requires prominent labeling of hazardous workplace and household products and chemicals. |
| | Multiple Use Act defines purposes for which national forests can be used. |
| 1963 | Clean Air Act Amendments extend research efforts to reduce vehicle emissions and establishes federal control over interstate air pollution; further amendments 1970,* 1977, 1990. |

1964        Wilderness Act.

            Land and Water Conservation Fund Act sets aside funding for
            local, state, and federal acquisition and development of land for
            parks and other open areas.

1965        Highway Beautification Act bans many types of billboards from
            highways.

            *California Land Conservation Act (Williamson Act).

            Water Quality Act gives the federal government power to enact
            water standards if state action is lacking.

1966        Clean Water Restoration Act expands and centralizes water pol-
            lution programs in Department of Interior; strengthens efforts in
            sewage treatment and water purification.

            National Historic Preservation Act.

1967        Air Quality Act establishes federal program of criteria, standards
            development, and enforcement of air pollution control.

1968        Wholesale Meat Act expands and updates Meat Inspection Act of
            1907.

            Wild and Scenic Rivers Act identifies areas of scenic beauty for
            the purpose of setting them aside for recreation and preservation.
            Aircraft Noise Abatement Act is the first federal effort to deal with
            health hazards of noise pollution.

1969        *National Environmental Policy Act (NEPA) creates the Council
            on Environmental Quality.

1970        Resource Recovery Act shifts focus of waste disposal from solid
            waste disposal to control, recovery, and recycling.

            Occupational Safety and Health Act (OSHA) establishes federal
            program to set standards for and enforce workplace safety and
            health.

            *Clean Air Act Amendments; further amendments 1977, 1990.

            Environmental Education Act ensures that teachers receive sup-
            plemental education in environmental issues; amended 1974.

1972        *Federal Water Pollution Control Act Amendments (popularly
            known as the Clean Water Act [CWA]) forbids discharge of pol-
            lutants into navigable waters; further amendments 1977, 1987.

            Marine Mammal Protection Act aims to protect and conserve ma-
            rine mammals and to encourage international research; amended
            1988.

Environmental Pesticide Control Act.

Coastal Zone Management Act.

Marine Protection, Research, and Sanctuaries Act (Ocean Dumping Act); amended 1977, 1988.

Federal Environmental Pesticide Control Act expands product registration, labeling, and environmental protection and expands monitoring of pesticide residues in food and water.
Noise Control Act.

1973          *Endangered Species Act; amended 1988.

1974          Safe Drinking Water Act (SDWA) requires the Environmental Protection Agency to set national drinking water standards; amended 1986.

1976          Toxic Substances Control Act (TOSCA).
              Fisheries Conservation and Management Act extends U.S. jurisdiction over marine fisheries resources to within 200 miles of coast and establishes regional fisheries councils made up of fishermen to regulate fishing activity.

              Federal Land Policy and Management Act regulates lands adjacent to federal lands.

1977          Surface Mining Control and Reclamation Act (Strip-Mining Act).

1978          Outer Continental Shelf Lands Act imposes liability for oil pollution caused by offshore operations.

1980          *Comprehensive Environmental Response, Compensation, and Liability Act (CERCLA) (Superfund Act) deals with the release of hazardous substances caused by spills or from abandoned dump sites and provides funds for cleanup of dump sites.
              Energy Security Act.

              Alaska National Interests Lands Act adds 103 million acres of park refuge and wilderness areas to U.S. protected land holdings.
              Act to Prevent Pollution from Ships.

1981          *Coastal Barrier Resources Act.

1984          Resource Conservation and Recovery Act (RCRA) attempts to make provision for the safe treatment and disposal of hazardous waste. It is the major federal law regulating solid waste disposal by municipalities.

1985          Swampbuster Provision of the Food Security Act (Farm Bill) discourages farmers from using converted wetlands for food production.

1987        Water Quality Control Act.

            Marine Plastic Pollution Research and Control Act (Annex V of International Convention for the Prevention of Pollution from Ships) prohibits dumping of plastics at sea.

1988        Ocean Dumping Ban Act (Title I of 1972 Marine Protection, Research, and Sanctuaries Act) bans dumping of industrial waste and sewage sludge into ocean.

1990        Oil Pollution Control Act defines liability for oil spills.

            Clean Air Act Amendments.

1994        California Desert Protection Act.

1996        Sustainable Fisheries Act requires the nation's eight regional fishing councils to adopt plans for fishery restoration and impose quotas where necessary.
            Antarctica Environmental Protection Act prohibits mining in Antarctica for at least fifty more years.

# Appendix II: Major International Agreements Relating to the Environment

This list includes bilateral and multilateral treaties, declarations, protocols, and conventions to which the United States is a party. The dates for the treaties and conventions are those when the agreements were opened for signature. Some of the agreements have not yet been ratified by the U.S. Senate, and a few are not yet in force. Items marked with an asterisk appear as documents in the text.

| | |
|---|---|
| 1916 | Migratory Bird Treaty between the United States and Canada. |
| 1940 | Convention on Nature Protection and Wildlife Preservation in the Western Hemisphere. |
| 1946 | International Convention for the Regulation of Whaling; amended 1956. |
| 1954 | Convention for the Prevention of Pollution of the Sea by Oil; amended 1969. |
| 1961 | Antarctic Treaty sets aside the continent for use "for peaceful purposes only" and provides a base for national groups to perform geophysical and biological research and freely exchange research data. |
| 1963 | Nuclear Test Ban Treaty among the United States, the United Kingdom, and the Soviet Union. |
| 1971 | Convention on Wetlands of International Importance, Especially as Waterfowl Habitat (Ramsar Convention); protocol added in 1982. |
| 1972 | *Stockholm Declaration on the Human Environment. |
| | Convention on the Prevention of Marine Pollution by Dumping of Waste and Other Matter (London Dumping Convention [LDC]). |
| | Convention on Biological Weapons bans germ warfare. |

1973        International Convention for the Prevention of Pollution from Ships (MARPOL); annexes added 1978; amended 1985.

            Convention on International Trade in Endangered Species of Wild Flora and Fauna (CITIES); amended 1979.

1979        Convention on Long-Range Transboundary Air Pollution (LRTAP) (Geneva Convention).

1980        U.S.-Canada Memorandum of Intent on Transboundary Air Pollution.

1982        *United Nations Convention on the Law of the Sea (UNCLOS) (governs ocean use and the exploitation of ocean resources). Not ratified by U.S. Senate.

1985        *Vienna Convention on Ozone Protection; *Montreal Protocol added in 1987.

1987        *Montreal Protocol on Substances That Deplete the Ozone.

1989        Basel Convention on the Control of Transboundary Movements of Hazardous Wastes and Their Disposal. Not ratified by U.S. Senate.

1991        United Nations moratorium on large drift nets (General Assembly action).

1992        *United Nations Framework Convention on Climate Change; Kyoto Protocol added in 1997.

            Convention on Biological Diversity. Not ratified by U.S. Senate.

1993        Chemical Weapons Treaty.

1996        Comprehensive Nuclear Test Ban Treaty (CTBT).

1997        *Kyoto Protocol on Climate Change. Not ratified by U.S. Senate.

# Notes

## INTRODUCTION

1. Robert Famighetti, ed., *World Almanac and Book of Facts*, 1999 (Mahwah, NJ: Primedia, 1998), p. 590. Other sources estimate that the world population reached 1 billion between 1775 and 1800.

2. John DeCicco, *$CO_2$ Diet for a Greenhouse Planet: A Citizen's Guide for Slowing Global Warming* (New York: Audubon Society, 1990), p. 8.

3. U.S. Bureau of the Census, "Energy Supply and Deposition, by Types of Fuel," *Statistical Abstract of the United States, 1998*, publication no. 948 (http://www.census.gov/statab/Freq/98s0948.txt).

## PART I

1. Estimates for the first human crossing of the Bering Strait range from as early as forty thousand years ago to as late as twelve thousand years ago (from which time there is actual evidence of human habitation). The scenario for the original settlement of the Americas that is offered here is based on recent anthropological, linguistic, and geological evidence supporting a comparatively early date for the first human infiltration of America. See, for example, Ann Gibbons, "Mother Tongues Trace Steps of Earliest Americans," *Science* 279 (February 27, 1998): 1306–7.

2. At the time of the Lewis and Clark Expedition, it was estimated that the Indian population of the West was about 225,000 and that of the Great Plains was about 150,000, and it seems reasonable to assume that, as a result of disease and warfare, the population had decreased by about one-third since the first contact between Europeans and Native Americans.

3. See St. Francis of Assisi, "The Song of Brother Sun and of All Creatures," in *St. Francis of Assisi: His Life and Writing as Recorded by His Contemporaries*, trans. Leo Sherley-Price (New York: Harper, 1960), pp. 161–62.

4. Thomas A. Bailey and David M. Kennedy, *The American Pageant*, 8th ed.,

Vol. 1, (Lexington, MA: Heath, 1987), p. 7. One of the earliest evaluations of the environmental impact of the European exploration and colonization of the Americas is to be found in Alfred W. Crosby, Jr., *The Columbian Exchange: Biological and Cultural Consequences of 1492* (Westport, CT: Greenwood Press, 1972).

5. Quoted in Marine Protection, Research, and Sanctuaries Act of 1972, P.L. 92-532, in *United States Statutes at Large*, Vol. 86 (Washington, D.C.: Government Printing Office, 1973), p. 4236.

6. John Reinhold Forster, Preface to Peter Kalm, *Travels into North America*, trans. by John Reinhold Forster, Vol. 1 (Warrington, England: William Eyres, 1772), pp. v–vi.

7. *Beowulf*, trans. Howell D. Chickering, Jr. (Garden City, NY: Anchor/Doubleday, 1977), p. 55.

8. www.AmPhilSoc.Org.

## PART II

1. Richard A. Bartlett, *The New Country: A Social History of the American Frontier, 1776–1890* (New York: Oxford University Press, 1974), p. 239, claims that John Quincy Adams "was the first President to manifest any real interest in conservation and was ahead of his time in comprehension of the problem." However, we found no corroboration of direct interest in conservation in either biographical or autobiographical works. Adams, as a prudent New Englander and former secretary of state, understood shipbuilding prerogatives and was simply being realistic about the needs of the navy and the New England fishing industry.

## PART III

1. J. J. Cosgrove, *History of Sanitation* (Pittsburgh: Standard Sanitary Manufacturing Co., 1909), p. 87.

2. See "Croton Aqueduct" and "Water," in Kenneth T. Jackson, ed., *The Encyclopedia of New York City* (New Haven, CT: Yale University Press, 1995), pp. 300–301, 1244–45.

3. Cosgrove, *History of Sanitation*, p. 88.

4. Haeckel, *Generelle Morphologie der Organismen*, trans. and quoted in Council of Learned Societies, *Dictionary of Scientific Biography*, Vol. 6 (New York: Scribner's, 1972), p. 8.

5. See Mark H. Brown, *The Plainsmen of the Yellowstone: A History of the Yellowstone Basin* (New York: Putnam's, 1961), pp. 188–94.

6. Now Section 1 of Article XIV of the New York State Constitution.

## PART IV

1. U.S. Bureau of the Census, *Historical Statistics of the United States* (Washington, D.C.: Government Printing Office, 1975). The 1850 figures actually apply just to Massachusetts, the only state for which figures are available.

2. Foreword to Frederick J. Turner, *The Frontier in American History* (N.P.: Readex, 1966), p. v.

3. Gifford Pinchot, *The Adirondack Spruce* (New York: The Critic, 1898), p. 110.

## PART V

1. George Bird Grinnell and Charles Sheldon, eds., *Hunting and Conservation* (New Haven, CT: Yale University Press, 1925), p. vii.

2. Henry Beston, *The Outermost House* (New York: Henry Holt, 1992), p. 10.

3. See, for example, the work of one of their chief researchers, Harold Barnett: Harold Barnett and Chandler Morse, *Scarcity and Growth: Economics of Natural Resource Availability* (Baltimore: Johns Hopkins University Press, 1963). The Galbraith selection, Document 94, comes from another publication prepared by Resources for the Future.

4. See, for example, Donella H. Meadows et al., *The Limits to Growth: A Report for the Club of Rome's Project on the Predicament of Mankind* (New York: Universe Books, 1972).

## PART VI

1. Edwin Dale, Jr., "The Economics of Pollution," *New York Times Magazine*, April 19, 1970, pp. 27–29, 40, 42, 44, 46.

2. Lynn White, Jr., "The Historical Roots of Our Environmental Crisis," *Science* 155, no. 3767 (March 10, 1967): 1206.

3. After more than $100 million had been spent and the project was in its final stages, the Supreme Court ruled that the dam could not be completed because the operation of the dam would destroy the only known habitat of the snail darter. See *Tennessee Valley Authority v. Hill*, in *United States Reports, 1978*, Vol. 437 (Washington, D.C.: Government Printing Office, 1980), pp. 153–213.

4. John H. Cushman, Jr., with Evelyn Nieves, "In Colorado Resort Fires, Culprits Defy Easy Labels," *New York Times*, October 24, 1998, p. A11.

5. Daniel P. Beard, "Dams Aren't Forever," *New York Times*, October 6, 1997, p. A19.

## PART VII

1. Quoted in Bill Hosakawa, "It Takes a Lot of Energy to Keep Up with Interior's Jim Watt," *Denver Post*, March 1, 1981, p. 25.

2. Gary Lucier, Susan Pollack, and Agnes Perez, "Import Penetration in the U.S. Fruit and Vegetable Industry," *Vegetables and Specialties*, VGS-273 (Washington, D.C.: Economic Research Service/USDA, November 1997), p. 16.

3. Mark Sagoff, "Do We Consume Too Much?" *Atlantic Monthly*, June 1977, p. 16.

4. World Commission on Environment and Development, *Our Common Future* (New York: Oxford University Press, 1987); and *Agenda 21 Earth Summit: The United Nations Programme of Action from Rio* (New York: United Nations, 1992).

5. Arne Naess, "The Shallow and the Deep, Long-Range Ecology Movements," in George Sessions, ed., *Deep Ecology for the Twenty-first Century: Readings in the Philosophy of the New Environmentalism* (Boston: Shambala, 1995), p. 151.

6. Edward O. Wilson, "Integrated Science and the Coming Century of the Environment," *Science* 279 (March 27, 1996): 2049.

7. Moustafa Tolba, "The Ozone Agreement—and Beyond," *Environmental Conservation* 14, no. 4 (1967): 287–90, quoted in Robert F. Fleagle, *Global Environmental Change: Interactions of Science, Policy and Politics in the United States* (Westport, CT: Praeger, 1994), p. 185.

8. U.S. Department of Health and Human Services, *Report on Black and Minority Health* (Washington, D.C.: U.S. Department of Health and Human Services, 1985), named for Health and Human Services secretary Margaret Heckler.

9. Intergovernmental Panel on Climate Change, Working Group I, *Climate Change: The IPCC Scientific Assessment*, ed. J. T. Houghton, G. J. Jenkins, and J. J. Ephraums (Cambridge: Cambridge University Press, 1990).

10. Bill McKibben, "The Earth Does a Slow Burn," *New York Times*, May 3, 1997, p. 23.

11. See, for example, the discussion in Theo Colborn, Dianne Dumanoski, and John Peterson Myers, *Our Stolen Future* (New York: Dutton, 1996), pp. 122–30, of Ana Soto and Carlos Sonnenschein's discovery of "hormone-disrupting chemicals where you would least expect them—in ubiquitous products considered benign and inert" (p. 125).

12. P. M. Vitousek, H. A. Mooney, J. Lubchenco, and J. M. Melillo, "Human Domination of Earth's Ecosystems," *Science* 277 (1997): 494.

# Glossary

**acid rain**   water in the atmosphere that, having picked up acids or acid-forming compounds (such as carbon dioxide) from the air, falls to the earth in the form of rain, snow, or sleet.

**ambient temperature**   temperature of the air in the surrounding space.

**anthropocentrism**   belief that human beings are the most important entities in the universe.

**barrier island**   ocean island that lies roughly parallel to the coast and acts as a protective barrier for the mainland against the ravages of the ocean.

**biocentrism**   belief that all living things have an equal right to flourish and enjoy the benefits of their common environment.

**biodiversity**   variation of the genetic material within a single species, of the totality of species within all taxonomic groupings, and of the landscapes and ecosystems within which these varied species exist.

**biosphere**   the segment of the earth where life is found.

**biota**   the totality of the flora and fauna found in a given area.

**bison**   American species of buffalo.

**carbon sink**   forest, deep ocean, or geological reservoir where carbon dioxide can be absorbed or sequestered away from the atmosphere.

**chlorofluorocarbons (CFCs)**   organic compounds, such as Freons, that contain chlorine, fluorine, and carbon atoms and, if they are gas, cause a depletion of atmospheric ozone when they are released into the air.

**clear-cutting**   cutting all the trees within a segment of a forest.

**climate change**   change in temperature or weather patterns (such as amount of rainfall) induced by human activity.

**commons**   resources that do not belong to any individual or nation, such as clean air, migratory birds, and noncoastal ocean resources, as well as property held in common by a community or nation, such as park land, communal grazing lands, and community dump sites.

**conservation**   natural resource management to prevent resource destruction, exploitation, and neglect.

**DDT**   dichlorodiphenyltrichloroethane, a chlorinated hydrocarbon (hydrogen-carbon compound) once widely used as a pesticide.

**desalination**   removal of dissolved salts from brackish or salty water to make the water usable for drinking or agriculture.

**drift net**   large fishing net designed to drift near the surface of the water and used by fishermen to trap fish or other marine animals.

**ecology**   the study of the relationships among living things and between living things and their nonliving environment.

**ecosphere**   segment of the universe inhabited by living things and where there is interaction among living things and between organisms and nonliving things.

**ecosystem**   ecological community of organisms and their environment that functions as a unit.

**endangered species**   species of wild flora or fauna whose numbers have been so reduced that it is in danger of becoming extinct.

**energy**   usable power (such as heat or electricity) or the resources to produce usable power (such as coal, water, or sunlight).

**entropy**   the increasing degradation of a society or system.

**environment**   totality of biotic, chemical, and physical elements (including air, climate, living things, and nonliving things) that affect an organism or a community.

**environmentalism**   advocacy of the conservation and maintenance of the natural environment and the limitation or reversal of human impact on it.

**environmental justice**   environmental actions based on ethical, moral, and spiritual factors rather than on economic interest or personal preference.

**estuary**   passageway at the lower part of a river where the sea tide meets the river.

**ethic**   value system that defines what is good and bad behavior and sets forth moral duties and obligations.

**fossil fuels**   coal, natural gas, petroleum, and other natural resources used as fuel.

**global warming**   climate change marked by a worldwide increase in average temperature.

**greenhouse effect**   warming of the lower layers of the atmosphere and of the earth's surface as a result of the increase of carbon dioxide and other heat-trapping gases in the atmosphere.

**greenhouse gases**   carbon dioxide, methane, and other gases that allow the sun's radiant energy to penetrate the earth's atmosphere and then prevent the resultant heat from escaping.

**green movements** wide-ranging social and political activities that focus on ecological issues.

**habitat** place where an organism or group of organisms lives.

**land reclamation** the process of making wetlands and arid land suitable for human uses, including agriculture and the construction of buildings.

**microenvironment** the total ecosystem of a small area, such as a pond, a forest, or a field.

**miner's canary** bird or other living thing whose activity can be used to monitor the presence of an environmental hazard.

**monoculture** farming, including tree farming, that involves raising only one crop or one species of plant over a wide area.

**natural resource** any material, either organic or inorganic, that is found in nature and is of use to humans.

**organism** any living thing, from a microbe to a tree to a human being.

**ozone depletion** destruction of the gaseous shield that protects the earth from harmful radiation.

**predation** the killing and consuming of animals of other species as a primary means of obtaining food.

**recycling** processing materials for reuse.

**renewable resources** energy sources, such as wind and sun, that are not depleted as a result of use, and living resources, such as plants and animals, that can be regenerated.

**restoration ecology** attempts to restore ecosystems to a natural state by eliminating introduced species and replacing them with species that at one time flourished in the region and by eliminating dams and other man-made structures.

**strip mining** mining in which power shovels, bulldozers, and other large machinery is used to remove huge chunks of the surface land.

**sustainable development** economic development, ranging from farming to industrial activity, that encourages the conservation, maintenance, and renewal of natural resources.

**watershed** region where numerous underground springs, small streams, and other small water courses deliver water to major rivers or large bodies of water, such as lakes and reservoirs.

**wetlands** swamps, marshes, bogs, tidal flats, permafrost, and other areas characterized by fluctuating levels of water and an abundance of plants, such as *Spartina*, that require a great deal of water.

**wilderness** an area that, together with the living things on it, has been fundamentally undisturbed by human activity.

**wildlife** nonhuman living things that have not been domesticated.

# Further Readings

Albright, Horace M. (as told to Robert Cahn). *The Birth of the National Park Service.* Salt Lake City: Howe Brothers, 1985.

Ambrose, Stephen E. *Undaunted Courage: Meriwether Lewis, Thomas Jefferson, and the Opening of the American West.* New York: Simon and Schuster, 1996.

Appleman, Philip, ed. *Thomas Robert Malthus: An Essay on the Principle of Population: Text, Sources and Background, Criticism.* New York: W. W. Norton, 1976.

Bartlett, Richard A. *The New Country: A Social History of the American Frontier, 1776–1890.* New York: Oxford University Press, 1974.

Beegel, Susan F., Susan Shillinglaw, and William N. Tiffney, Jr., eds. *Steinbeck and the Environment.* Tuscaloosa: University of Alabama Press, 1997.

Benedick, Richard. *Ozone Diplomacy: New Directions in Safeguarding the Planet.* Cambridge, MA: Harvard University Press, 1991.

Berry, Thomas. *The Dream of Earth.* San Francisco: Sierra Club Books, 1988.

Berry, Wendell. *Home Economics: Fourteen Essays.* San Francisco: North Point Press, 1987.

Billington, Ray Allan. *Land of Savagery, Land of Promise.* New York: W. W. Norton, 1986.

Blake, Nelson M. *Land into Water—Water into Land: A History of Water Management in Florida.* Tallahassee: University Presses of Florida, 1980.

Boyle, Robert H. *The Hudson River: A Natural and Unnatural History.* New York: W. W. Norton, 1979.

Brown, Harrison. *The Human Future Revisited: The World Predicament and Possible Solutions.* New York: W. W. Norton, 1978.

Brown, Lester R. *Who Will Feed China?* New York: W. W. Norton, 1995.

Buchanan, James M. *The Limits of Liberty: Between Anarchy and Leviathan.* Chicago: University of Chicago Press, 1975.

Cahn, Robert. *Footprints on the Planet: A Search for an Environmental Ethic.* New York: Universe Books, 1978.

Carter, Luther K. *The Florida Experience: Land and Water Policy in a Growth State.* Baltimore: Johns Hopkins University Press, 1974.

Cohen, Michael P. *History of the Sierra Club, 1892–1970.* San Francisco: Sierra Club Books, 1988.

Colborn, Theo, Dianne Dumanoski, and John Peterson Myers. *Our Stolen Future.* New York: Dutton, 1996.

Coyle, David Cushman. *Conservation: An American Story of Conflict and Accomplishment.* New Brunswick, NJ: Rutgers University Press, 1957.

Cronon, William. *Nature's Metropolis: Chicago and the Great West.* New York: W. W. Norton, 1991.

———. *Changes in the Land: Indians, Colonists and the Ecology of New England.* New York: Hill and Wang, 1983.

Crosby, Alfred W., Jr. *The Columbian Exchange: Biological and Cultural Consequences of 1492.* Westport, CT: Greenwood Press, 1972.

———. *Ecological Imperialism: The Biological Expansion of Europe.* New York: Cambridge University Press, 1986.

Daily, Gretchen C., ed. *Nature's Services: Society's Dependence on Natural Ecosystems.* Washington, D.C.: Island Press, 1997.

Dale, Edwin, Jr. "The Economics of Pollution." *New York Times Magazine,* April 19, 1970, pp. 27–28, 40–46.

Daly, Herman. *Toward a Steady State Economy.* San Francisco: W. H. Freeman, 1973.

Darling, Jay N. *John Muir and His Legacy.* Boston: Little, Brown, 1981.

Dean, Cornelia. *Against the Tide: The Battle for America's Beaches.* New York: Columbia University Press, 1999.

De Bry, Theodore. *Discovering the New World.* Ed. Michael Alexander. New York: Harper and Row, 1976.

de Steiguer, J. E. *Age of Environmentalism.* New York: McGraw-Hill, 1997.

Devall, Bill, and George Sessions. *Deep Ecology: Living as if Nature Mattered.* Salt Lake City: Gibbs Smith, 1985.

Duncan, Dayton, and Ken Burns. *Lewis and Clark: The Journey of the Corps of Discovery.* New York: Knopf, 1997.

Ehrlich, Paul, and Anne Ehrlich. *Population, Resources, Environment: Issues in Human Ecology.* San Francisco: W. H. Freeman, 1970.

Engle, J. Ronald. *Sacred Sands: The Search for Community.* Middletown, CT: Wesleyan University Press, 1983.

Esterbrook, Gregg. *A Moment on Earth: The Coming of Age of Environmental Optimism.* Penguin, 1995.

Fleagle, Robert G. *Global Environmental Change: Interactions of Science, Policy, and Politics in the United States.* Westport, CT: Praeger, 1994.

Fox, Stephen R. *John Muir and His Legacy: The American Conservation Movement.* Boston: Little, Brown, 1981.

Franck, Irene, and David Brownstone. *The Green Encyclopedia.* New York: Prentice-Hall, 1992.

Garrett, Laurie. *The Coming Plague: Newly Emerging Diseases in a World Out of Balance.* New York: Farrar, Straus and Giroux, 1994.

Geller, Lawrence D., ed. *Pilgrims in Eden: Conservation Policies at New Plymouth.* Wakefield, MA: Pride Publications, 1974.

Glickson, Artur. *The Ecological Basis of Planning*. Ed. by Lewis Mumford. The Hague: Nijhoff, 1971.

Gottlieb, Allan. *The Wise Use Agenda*. Bellevue, WA: Free Enterprise, 1989.

Graham, Frank, Jr. *The Audubon Ark*. Austin: University of Texas Press, 1992.

Grinnell, George Bird. *The Passing of the Great West: Selected Papers of George Bird Grinnell*. Ed. by John F. Reiger. New York: Winchester, 1972.

Hays, Samuel P. *Beauty, Health, and Permanence: Environmental Politics in the United States, 1958–1985*. New York: Cambridge University Press, 1987.

Helvarg, David. *The War Against the Greens: The "Wise-Use" Movement, the New Right and Anti-environmental Violence*. San Francisco: Sierra Club Books, 1994.

Hunter, J. Robert. *Simple Things Won't Save the Earth*. Austin: University of Texas Press, 1997.

Kamen, David. *Song of the Dodo*. New York: Scribner's, 1996.

Keeble, John. *Out of the Channel: The* Exxon Valdez *Oil Spill in Prince William Sound*. New York: HarperCollins, 1991.

Krech, William, III. *The Ecological Indian: Myth and History*. New York: W.W. Norton, 1999.

Lappe, Frances Moor. *Diet for a Small Planet*. New York: Random House/Ballantine, 1992.

Lear, Linda. *Rachel Carson: Witness for Nature*. New York: Holt, 1997.

Leiss, William. *The Domination of Nature*. New York: Braziller, 1972.

Levine, Adeline Gordon. *Love Canal: Science, Politics, and People*. Lexington, MA: Lexington Books, 1982.

Linton, R. N. *Terracide*. Boston: Little, Brown, 1970.

List, Peter C. *Radical Environmentalism: Philosophy and Tactics*. Belmont, CA: Wadsworth, 1993.

Lopez, Barry. *Of Wolves and Men*. New York: Macmillan, 1978.

McHenry, Robert, and Charles Van Doren, eds. *A Documentary History of Conservation in America*. New York: Praeger, 1972.

McKibben, Bill. *The End of Nature*. New York: Anchor/Doubleday, 1989.

McPhee, John. *The Control of Nature*. New York: Farrar, Straus and Giroux, 1989.

————. *Encounters with the Archdruid*. New York: Noonday Press, 1971.

Margolis, Howard. *Dealing with Risk: Why the Public and the Experts Disagree on Environmental Issues*. Chicago: University of Chicago Press, 1990.

Marx, Leo. *The Machine in the Garden: Technology and the Pastoral Idea in America*. New York: Oxford University Press, 1964.

Merchant, Carolyn. *Major Problems in American Environmental History*. Lexington, MA: Heath, 1993.

Morgan, Arthur E. *Dams and Other Disasters: A Century of the Army Corps of Engineers in Civil Work*. Boston: Porter Sargent, 1971.

Muir, John. *Our National Parks*. New York: AMS Press, 1970. Reprint of 1901 ed.

Myers, Norman. *The Sinking Ark*. Oxford, England: Pergamon Press, 1979.

Nash, Roderick Frazier. *The Rights of Nature*. Madison: University of Wisconsin Press, 1989.

————. *Wilderness and the American Mind*. 3rd ed. New Haven: Yale University Press, 1982.

————, ed. *Readings in the History of Conservation*. Reading, MA: Addison-Wesley, 1968.

Olmsted, Frederick Law, Jr., and Theodora Kimball, eds. *Frederick Law Olmsted: Landscape Architect, 1822–1903*. Bronx, NY: Benjamin Blom, 1970. Reprint of *Forty Years of Landscape Architecture: Being the Professional Papers of Frederick Law Olmsted, Senior* (1928).

Passmore, John Arthur. *Man's Responsibility for Nature: Ecological Problems and Western Traditions*. London: Duckworth, 1974.

Petulla, Joseph M. *American Environmental History*. 2nd ed. Columbus, OH: Merrill, 1988.

Powell, John Wesley. *Selected Prose of John Wesley Powell*. Ed. by George Crossette. Boston: Godine, 1970.

President's Council on Sustainable Development. *Sustainable America: A New Consensus for Prosperity, Opportunity, and a Healthy Environment*. Washington, D.C.: President's Council on Sustainable Development, 1996.

Reich, Charles A. *The Greening of America: How the Youth Revolution Is Trying to Make America Livable*. New York: Random House, 1970.

Reiger, John F. *American Sportsmen and the Origins of Conservation*. New York: Winchester Press, 1975.

Repetto, Robert. *Wasting Assets: The Need for Natural Resources Accounting*. Washington, D.C.: World Resources Institute, 1989.

Rhodes, Richard. *The Making of the Atomic Bomb*. New York: Simon and Schuster, 1986.

Rogers, Marion Lane, ed. *Acorn Days: The Environmental Defense Fund and How It Grew*. New York: Environmental Defense Fund, 1990.

Roosevelt, Franklin D. *Franklin D. Roosevelt and Conservation, 1911–1945*. Ed. by Edgar B. Nixon. 2 vols. Hyde Park, NY: FDR Library, 1957.

Roosevelt, Theodore. *An Autobiography*. New York: Macmillan, 1914.

Rubin, Charles T. *The Green Crusade: Rethinking the Roots of Environmentalism*. New York: Free Press, 1994.

Rybczynski, Witold. *A Clearing in the Distance: Frederick Law Olmsted and America in the Nineteenth Century*. New York: Scribner's, 1999.

Safina, Carl. *Song for the Blue Ocean: Encounters along the World's Coasts and beneath the Sea*. New York: Holt, 1997.

Sagoff, Mark. "Do We Consume Too Much?" *Atlantic Monthly*, June 1997, pp. 80–96.

Sale, Kirkpatrick. *The Green Revolution: The American Environmental Movement, 1962–1992*. New York: Hill and Wang, 1993.

Sanger, Marjory Bartlett. *Billy Bartram and His Green World: An Interpretive Biography*. New York: Farrar, Straus and Giroux, 1972.

Schneider, Paul. *The Adirondacks*. New York: Henry Holt, 1997.

Sessions, George, ed. *Deep Ecology for the Twenty-first Century: Readings on the Philosophy of the New Environmentalism*. Boston: Shambala, 1995.

Shabecoff, Philip. *A Fierce Green Fire: The American Environmental Movement*. New York: Hill and Wang, 1993.

Shanks, Bernard. *This Land Is Your Land: The Struggle to Save America's Public Lands*. San Francisco: Sierra Club, 1982.

Squire, C. B. *Heroes of Conservation*. New York: Fleet, 1978.

Strong, Douglas H. *Dreamers and Defenders: American Conservationists*. Lincoln: University of Nebraska Press, 1971.

Strasser, Susan. *Waste and Want: A Social History of Trash*. New York: Holt/ Metropolitan Books, 1999.

Thorne-Miller, Boyce, and John Catena. *The Living Ocean: Understanding and Protecting Marine Biodiversity*. Washington, D.C.: Island Press, 1991.

Tobey, Ronald C. *Saving the Prairies: The Life Cycle of the Founding School of American Plant Ecology, 1895–1955*. Berkeley: University of California Press, 1981.

Tokar, Brian. *The Green Alternative: Creating an Ecological Future*. San Pedro, CA: Miles, 1987.

U.S. Department of Health, Education and Welfare, *Health in America: 1776–1976*. Publication no. HRA 76–616. Washington, D.C.: Government Printing Office, 1976.

Ward, Barbara, and Rene Dubos. *Only One Earth: The Care and Maintenance of a Small Planet*. New York: W. W. Norton, 1972.

Warren, Louis S. *The Hunter's Game: Poachers and Conservationists in Twentieth-Century America*. New Haven, CT: Yale University Press, 1997.

Watkins, T. H. "Untrammeled by Man: The Making of the Wilderness Act of 1964." *Audubon*, November 1984, pp. 74–90.

Wilkinson, Richard. *Progress and Poverty*. New York: Praeger, 1973.

Wilson, Edward O. *Naturalist*. New York: Warner Books, 1994.

World Commission on Environment and Development. *Our Common Future*. New York: Oxford University Press, 1987.

Worldwatch Institute. *State of the World 1993*. New York: W. W. Norton, 1993.

Worster, Donald. *Nature's Economy: A History of Ecological Ideas*. New York: Cambridge University Press, 1977.

———. *The Wealth of Nature: Environmental History and the Ecological Imagination*. Oxford: Oxford University Press, 1993.

# Index

of standing bodies of, 89; management of, 126, 134–36; plants growing in, 204; policy, 161–62; pollution of, 7, 13, 76–77, 144, 185, 215, 217, 218, 239, 257, 262; pollution control legislation, 145; as source of energy, 11, 81, 128; standards for, 232; supply and supply systems, 7, 9–10, 73, 76–77, 89, 105, 114, 117, 120, 124, 128, 132, 143, 173, 218; usage of, 9–10, 76–77, 137, 164, 173, 253, 271; wastage of, 137, 144. *See also* hydroelectric power; irrigation; rivers and streams; waterways

watersheds, 10, 178
waterways, 52, 129
Watt, James, 134, 231
West: development of, 113–14, 134–35; expansion into, 52, 75, 77–78, 102, 121–22; population of, 20, 120, 124
wetlands, 47–49, 81, 87–88, 203, 217; control of growth near, 149; draining of, 165, 252; policy, 146; reclaiming of, 9, 48, 81, 88, 252–54. *See also* Everglades; marshes
White, Lynn, Jr., 198–99
wilderness, 11, 80, 177–78; appreciation and love of, 54, 79, 84, 122, 123;

attitudes toward, 21–22, 25, 43; importance of, 177–78; loss of, 144; Pilgrims and, 31–33; preservation of, 9, 71–72, 96; usefulness of, 218
Wilderness Act, 107, 145
Wilderness Society, 107, 145, 177
wildlife, 22; areas for, 218; disappearance and destruction of, 7, 54, 80, 144, 145, 258; laws to protect, 69, 70; management of, 245, 254; preservation of, 223, 224. *See also* habitat
Wilson, Edward, 233, 254–56, 271
wise use, 120, 124, 136–37, 231, 233, 242
wolves, 36, 37, 68, 112
wood, 15, 225. *See also* forests; timber
working conditions, 118, 126–27
World War I, 120, 143
World War II, 33, 143, 144, 146, 231, 232, 243
World Watch Institute, 242
Wyoming, 101, 125, 143

Yellowstone National Park, 80, 101–2
Yosemite National Park and Yosemite Valley, 80, 95–96, 132–33

zoning, 149–52, 160

**About the Editors**

PENINAH NEIMARK is a freelance editor and writer specializing in environment and development issues.

PETER RHOADES MOTT teaches biology and ethics at the Fieldston School in Riverdale, NY. He is President of the Board of the New York City chapter of the Audubon Society.

**Primary Documents in American History and Contemporary Issues**

The Abortion Controversy
*Eva R. Rubin, editor*

The AIDS Crisis
*Douglas A. Feldman and Julia Wang Miller, editors*

Capital Punishment in the United States
*Bryan Vila and Cynthia Morris, editors*

Constitutional Debates on Freedom of Religion
*John J. Patrick and Gerald P. Long, editors*

Founding the Republic
*John J. Patrick, editor*

Free Expression in America
*Sheila Suess Kennedy, editor*

Genetic Engineering
*Thomas A. Shannon, editor*

The Gun Control Debate
*Marjolijn Bijlefeld, editor*

Major Crises in Contemporary American Foreign Policy
*Russell D. Buhite, editor*

The Right to Die Debate
*Marjorie B. Zucker, editor*

The Role of Police in American Society
*Bryan Vila and Cynthia Morris, editors*

Sexual Harassment in America
*Laura W. Stein*

States' Rights and American Federalism
*Frederick D. Drake and Lynn R. Nelson, editors*

U.S. Immigration and Naturalization Laws and Issues
*Michael LeMay and Robert Barkan, editors*

Women's Rights in the United States
*Winston E. Langley and Vivian C. Fox, editors*